the gecko's foot

the gecko's foot

Bio-inspiration: Engineering New Materials from Nature

Peter Forbes

W. W. Norton & Company
New York · London

First American Edition 2006

First published in Great Britain in 2005 by Fourth Estate, an imprint of HarperCollinsPublishers, under the title *The Gecko's Foot: Bio-inspiration: Engineered from Nature.*

Manufacturing by Quebecor World, Fairfield
Production manager: Anna Oler

Library of Congress Cataloging-in-Publication Data

Forbes, Peter.
The gecko's foot : bio-inspiration : engineering new materials from nature / Peter Forbes — 1st American ed.
p. cm.
Includes bibliographical references and index.
ISBN-13: 978-0-393-06223-6
ISBN-10: 0-393-06223-6
1. Technological innovations. 2. Nature. I. Title.
T173.8.F63 2006
600—dc22 2006006731

W. W. Norton & Company, Inc.
500 Fifth Avenue, New York, N.Y. 10110
www.wwnorton.com

W. W. Norton & Company Ltd.
Castle House, 75/76 Wells Street, London W1T 3QT

1 2 3 4 5 6 7 8 9 0

In memory of my father
Leonard Harry Forbes (1916–1991)

CONTENTS

LIST OF ILLUSTRATIONS

ACKNOWLEDGMENTS

For this book, I interviewed many of the key practitioners of bio-inspiration and corresponded with many more. I am very grateful to those scientists, engineers and artists who made time to see me, answered my innumerable e-mail queries, made their illustrations available and checked facts in the typescript, especially Joanna Aizenberg, Steven Arcidiacono, Kellar Autumn, Julia Barfield, Wilhelm Barthlott, Angela Belcher, Paul Calvert, David Campbell, Tony Copeland, Charles Ellington, Ron Fearing, Robert Full, Andre Geim, Helen Ghiradella, Simon Hurst, Heinz Isler, David Kaplan, David Knight, Biruta Kresling, L Mahadevan, Stephen Mann, David Marks, George McDonald, Koryo Miura, Alex Parfitt, Tony Robbin, Kevin Sanderson, Mehmet Sarikaya, Kenneth Snelson, Julian Vincent, Christopher Viney, Pete Vukusic, Robin Wootton, and Rafał Żbikowski.

Many others helped with specific enquiries, especially Dietrich Bechert, Andrea Born, Yerev Braun, Reinhard Budde, Peter Butcher, John Chilton, Steve Chilton, Timothy Deming, Julian Ellis, Alastair Fowler, Laura Giuffrida, Simon Guest, Gordon Hendler, Donald Ingber, Holger Krapp, Randy Lewis, Adrian Marshall, Richard Milner, Carlo Montemagno, Dan Morse, Kenkichi Nose, Geoffrey Ozin, Anne Peattie, Sergio Pellegrino, Mieke van der Leeden, Fritz Vollrath, Binquing Wei, and Eli Yablonovitch.

Much of the research was carried out in the British Library at St Pancras, where pretty well all the world's scientific papers, patents and general books can now be made available in one room. I am especially grateful to Tim Radford, Science Editor of the *Guardian*, for encouragement over many years and for commissioning me to write a series of articles on bio-inspiration from 2000–3; this research formed the platform for the book. The book could never have been published without the shaping hand in the early stages of my agent Andrew Lownie; I am grateful to my initial editors at Fourth Estate, Nick Davies and Christopher Potter who believed in the idea and took the book on. Mitzi Angel, who became the book's editor at a crucial stage, had a powerful influence on the final shaping of the text. The book was completed during a Royal Literary Fund Fellowship at Queen Mary University of London. I am grateful to the RLF and Queen Mary for easing the burden. My partner, Diana Reich, heroically read the entire text and steered me away from many a textual blind alley.

For permission to reproduce the following extracts I am grateful to the following: to Random House Group Ltd for an extract from *Silk* by Alessandro Baricco, Harvill. Reprinted by permission of the Random House Group Ltd; to the Wylie Agency Inc for an extract from *Six Memos for the Next Millennium* by Italo Calvino (Cape, 1992) © 1992 by Italo Calvino, reprinted by permission of the Wylie Agency Inc.; to Thomson Publishing Services for an extract from *Adaptive Coloration in Animals* by H. B. Cott (Oxford University Press, 1940); to Liveright Publishing Corporation for an extract from *The Bridge* by Hart Crane (Liveright, 1992); to HarperCollins Publishers, Inc. for an extract from *Prey* by Michael Crichton (HarperCollins, 2002) © Michael Crichton (2002); to Chris Phoenix and Eric Drexler for an extract from 'Safe exponential manufacturing' by Eric Drexler and Chris Phoenix from *Nanotechnology*, 15, 2004; to California Institute of Technology for an extract from 'There's Plenty of Room at the Bottom', by Richard Feynman, from *Engineering and Science*, Vol 23, No 5, February, 1960, pp.22-36; to Helen Ghiradella for an extract from 'Light and Color on the Wing', *Applied Optics*, Vol 30 No 24, 1991; to David Higham Associates for an extract from *The Third Man* by Graham Greene (Heinemann, 1976); to HarperCollins Publishers Ltd for an extract from *The Open Sea: The World of Plankton* by Sir Alistair Hardy (Collins, 1956) © 1956 Sir Alistair Hardy; to Farrar, Straus and Giroux, L.L.C for "The

Bridge' from 'Ten Glosses' from *Electric Light* by Seamus Heaney. Copyright © Seamus Heaney. Reprinted by permission of Farrar, Straus and Giroux, L.L.C; to Bloodaxe Books Ltd for an extract from 'Brief reflection on cats growing on trees', by Miroslav Holub, translated by Ewald Osers from *Poems: Before & After: Collected English Translations* (Bloodaxe, 1990); to Simon & Schuster Adult Publishing Group for extracts from *Other People's Trades* by Primo Levi, translated by Raymond Rosenthal, Copyright © 1989 by Simon & Schuster. Reprinted with permission of Simon & Schuster Adult Publishing Group. All rights reserved; to Indiana University Press for an extract from *Lucretius: The Way Things Are*, translated by Rolf Humphries (Indiana University Press, 1969); to Random House Inc. for an extract from *Dr Faustus* by Thomas Mann, translated by John E. Woods (Vintage, 1999); to HarperCollins Publishers for an extract from 'Brooklyn Bridge' by Vladimir Mayakovsky, translated by Vladimir Markov and Merrill Sparks, from *Modern Russian Poetry* (MacGibbon & Kee, 1968); to Seren Books for an extract from 'Bumblebees and the Scientific Method', from *Id's Hospit* by Sheenagh Pugh (Seren Books, 1997); to Bloodaxe Books Ltd for an extract from 'The Spirit is too Blunt an Instrument' by Anne Stevenson, from *The Collected Poems 1955-1995* (Bloodaxe, 2000); to The MIT Press for an extract from *The Simple Science of Flight* by Henk Tennekes (MIT Press, 1996); to Cambridge University Press for extracts from D'Arcy Wentworth Thompson, *On Growth and Form*, Abridged Edition, edited by J. T. Bonner (Cambridge University Press, 1961); to Oxford University Press for an extract from *D'Arcy Wentworth Thompson: the scholar-naturalist, 1860-1948*, by Ruth D'Arcy Thompson (OUP, 1958). By permission of Oxford University Press; to The Consumers' Association for an extract from *Which?*, June 2003.

Every effort has been made to trace copyright holders of the extracts published in this book. The editor and publishers apologise if any material has been included without permission or without the appropriate acknowledgement, and would be glad to be told of anyone who has not been consulted.

CHAPTER ONE

Something New Under the Sun

These *Atom-Worlds* found out, I would despise
Colombus, and his vast Discoveries.

RICHARD LEIGH (1649–1728), 'Greatness in Little'

'Nature' is one of our great good words. To do things naturally, to go with the flow, to feel that we are in harmony with the principle that has sustained life on the planet for, according to our best guesses, more than three and a half billion years: all of these are natural (that word again) aspirations. But when we think of how we actually live – by means of technology – we feel 'unnatural': all our activities seem to involve forcing nature to do things she would otherwise not have done. We fear that perhaps we are a rogue species: the first one to have broken the bounds of nature.

These psychological feelings may or may not reflect the reality of our situation but there is no doubt that our technology and nature's are radically different. Our planes do not fly like birds and insects; although we travel faster than a cheetah, by muscle power alone we are much slower.

Many scientists now believe that it is possible for us to close the gap between our technology and nature. Bio-inspiration is the new science that seeks to use nature's principles to create things that evolution never achieved. To do this has entailed understanding nature at a new level – a tiny realm, far beneath our vision, and beneath the threshold of even the best optical microscopes.

Throughout human history human beings have been prejudiced creatures, and perhaps we were once biologically programmed to be that way. Despite this, we have learnt to cast aside narrow chauvinisms one by one and to embrace a broader view of our place in the scheme of things. But one set of blinkers remains: as adults we are creatures of a certain dimension – mostly 1.5–1.8 m tall – and we cannot help seeing things much smaller or larger than ourselves as remote from our experience. Apparently, we are deeply and stubbornly sizist.

The general acceptance, from the 17th century on, that the Earth was merely a planet of the Sun was supposed to have humbled our human pretensions. And the subsequent awareness of the vast distances of the universe, the number of stars (and, potentially, planets) and the minor-star status of the Sun were supposed to have increased this humiliation. The truth is, it is the things nearest to us that matter most. When we are ill in bed with flu, our horizon shrinks to our own body. And when we are bounding with health, it is pleasure on our own scale that we chase after. The universe can go run itself.

But this book is mostly about small things, not large, and they often seem even more distant than the black holes and supernovae of the deep universe. We find it quite hard to understand that minute creatures such as fleas and midges are fully functional, with a nervous system, a brain, a heart, and all the apparatus of life. In fact, life begins way below the threshold of human vision, and the intricately structured apparatus on which life depends – DNA, proteins and countless other molecules – is much smaller still.

For most of human history we have fabricated the devices we need on our own scale from simple materials, especially the metals such as iron, copper, zinc and tin. These are chemical elements and they are the same stuff all the way through – billions of atoms packed together like snooker balls in a frame, and then another layer on top, and so on *ad infinitum*. Biological materials, such as wood and cotton, have a much more complicated structure than metals and the intimate molecular structure of these materials was unknown until the 20th century. They were presented to us, more or less ready to use, and we used them without knowing what they were made from.

The microscope and telescope were both invented in the 17th century but it was the telescope that made the most impact. The telescope was always trained on some big new frontier – bigger ships, bigger factories, bigger armies – so it was something of a shock when the celebrated physicist Richard Feynman, in a talk of characteristic bravado given to the American Physical Society in 1959, announced that 'There's plenty of room at the bottom'. By this he meant that even as we ran out of personal space in our human-scale world, there was a paradoxically spacious untapped domain in which our minds could roam, one that was beneath the threshold of our vision. This was the nanorealm, in which objects are between one billionth and one millionth of a metre in size. Feynman suggested that this realm had room enough to do many things of great interest, and that life was already doing them, if only we could see what was going on:

> This fact … that enormous amounts of information can be carried in an exceedingly small space … is, of course, well known to the biologists … All this information … whether we have brown eyes, or whether we think at all, or that in the embryo the jawbone should first develop with a little hole in the side so that later a nerve can grow through it … all this information is contained in a very tiny fraction of the cell in the form of long-chain DNA molecules in which approximately 50 atoms are used for one bit of information about the cell.
>
> It is very easy to answer many of the fundamental biological questions; you just *look at the thing!* You will see the order of bases in the chain; you will see the structure of the microsome. Unfortunately, the present microscope sees at a scale which is just a bit too crude. Make the microscope one hundred times more powerful, and many problems of biology would be made very much easier.

It must have seemed crazy to many at the time. Feynman blithely asserted that the whole of the *Encyclopaedia Britannica* could be stored on the head of a pin. Now we can believe this because even if we have not quite got it down to a pinhead, we are not far off with our electronic disk-storage systems. But the micro-electronics revolution was only the first stage of the drive into micro-space. At the time, Feynman looked to biology to make his point because he knew that nature did her most intricate work on a tiny scale. But he also knew that most of the detail was tantalizingly out of reach.

A gecko climbs a vertical glass wall sure-footedly; when it reaches the ceiling it steps onto it and continues, upside-down, without difficulty. From the other side of the glass you can see transverse bands of tissues crossing its feet that alternately grip and release in a mini Mexican wave across the surface of the foot. A leaf of the sacred lotus unfurls in muddy water; as it rises, all the mud rolls off as if magnetically repelled, leaving a pristine surface. From a quarter of a mile away you can see the brilliant blue wings of a *Morpho rhetenor* butterfly; they are not just blue – they shimmer with an iridescent sparkle – but analysis reveals no blue pigment in the wings. That same *Morpho* butterfly takes off and jinks through the air, changing direction abruptly; until 1996, scientists were at a loss to understand how insects like this could fly. According to the well-tried aerodynamic theories that took a Jumbo into the air or flew Concorde at twice the speed of sound, insects did not generate enough lift to fly, but fly they do. And when a heavy insect thuds into a spider's web constructed from filaments about one tenth the diameter of human hair, the web distorts, brings the fly to a standstill and then returns to its original shape, the fly held fast in its sticky capture threads. Human engineering suggests that even if such a gossamer structure could catch an insect, it ought to fling it out again in recoil.

These creatures obviously possess skills and attributes beyond conventional engineering. But if we could find out how they achieve what they do, and learn how to utilize their techniques, it would extend our capabilities unimaginably. But the mechanisms behind these feats were hidden in structures so tiny that no microscope could observe them, and their chemical structures were so complex they defeated all attempts at analysis. As for creating man-made substances with the same properties: it was out of the question.

The dramatic powers of adhesion, self-cleaning, optical wizardry, tough elasticity and aerodynamics shown by these creatures are all highly prized by technologists. Scientists have long admired nature's engineering skills. Indeed, the precision of some of nature's gadgets takes the breath away: the stinging cells of jellyfishes; the jet engines of squids and cuttlefish; the marine creatures (and the land-based fireflies) that produce light without any heat. But there was no simple way of translating natural mechanisms into technical equivalents.

Nature was thought to use an entirely different set of principles to those of the engineer. Nature was soft and wet, worked at room temperature, and made her gadgets out of incredibly complex substances. While the human engineer instinctively reaches for metals to heat and beat into shape, nature goes for proteins that are grown inside living cells at body temperature. A single protein molecule is made from hundreds or thousands of smaller component molecules, virtually all of which have to be in precisely the right place for the protein to work.* A protein molecule is first made as a long chain and then it folds up precisely into a three-dimensional ball, like a piece of wet origami.

Nanotechnology has brought nature and engineering far closer together. If Feynman's 1959 talk is seen as the beginning of nano-technology, natural mechanisms were taken to be the epitome of the science right from the beginning. And now we don't just stare at creatures in amazement, wondering 'How do they do it?' Thanks to genetic engineering and a host of new techniques, we can now start to unravel nature's nanoengineering and produce engineered equivalents for it. This is bio-inspiration.

What makes bio-inspiration possible is the miracle that nature's mechanisms do not have to be 'alive' to work. In the 19th century, there was a doctrine known as 'vitalism' which held that all living things had a magical property – the *élan vital* – that could not be reduced to material science. Even the waste products of living things were thought to be fundamentally different from mineral substances. The doctrine began to crumble in 1828 when the waste product urea was made in the laboratory from two ordinary

* A few substitutions are possible in some proteins without impairing their action. For example, before insulin for diabetics could be genetically engineered, it was extracted from cattle, sheep or pigs, all of which have a slightly different composition to human insulin. In biological action, though, they are virtually identical.

chemicals of mineral origin. Thereafter, the idea of vitalism suffered blow after blow and now no scientist seriously believes that living things are, in a material sense, any more than the chemicals that comprise them. The property of life derives from the enormous complexity of the way the chemicals are organized, and not from an *élan vital*; some of the principles of this organization will become clear as the book proceeds.

Many of nature's most ingenious systems can continue to work outside living cells, in a test tube, and can be directed to work in novel ways to suit our purposes. For instance, in 1997 it was discovered that, although proteins will never meet such substances in the living cell, in the laboratory they can bind to inorganic materials such as gold and silver. Not only that but new proteins can be engineered that can bind to all the materials used to make computer chips. And since proteins are structured on a much smaller scale than silicon chips, they could act as templates for smaller microchips – nanochips.

Proteins have active centres, nooks and crannies precisely fashioned so that only one specific chemical can fit into them. When, in the whirling fluids of the cell, the one and only right chemical happens to come along, it becomes tightly bound to the protein. In living cells, proteins bind some chemicals, let others pass through pores, and, in general, regulate the traffic within the cell and facilitate chemical reactions. The full implications of this are spelt out in Chapter 6 but for now the point is that we have come so far from vitalism that the old division between living and non-living substances is breaking down – we can engineer hybrids between the two.

That there are no new frontiers is a weary cliché of our time: the ancient thrill of unspoiled places on Earth has given way to the fact of life that people can and do fly anywhere anytime. The dream of new worlds in space has retreated in the face of the barrenness of the Moon and Mars; the glorious new dawn of modernism in the Arts in the early 20th century led only to the stylistic emporium of post-modernism in which any retro style could be taken up again for a few years, given a whirl, then dropped. The decadence and satiation of our world is only too apparent. Scientifically, we have gone very deep

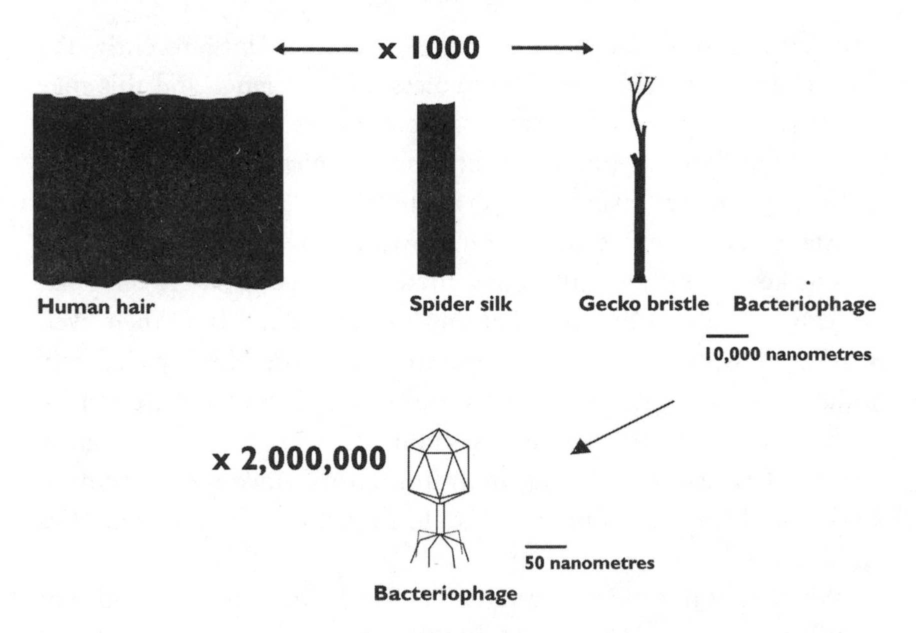

Fig. 1.1 'Plenty of room at the bottom'. Nature builds on a tiny scale. The bacteriophage, a virus that preys on bacteria, is an exquisite piece of engineering on a small scale and has an important role in bio-inspiration. But it is too small to be seen at the same degree of magnification as spider silk and gecko bristles, themselves very fine-scale structures compared to human hair. Nature does much of her most interesting work in the range 1–1,000 nanometres: a nanometre is one billionth of a metre (one millionth of a millimetre).

– into the nucleus of the atom and the genetic code of all life – so what can be left to discover?

Bio-inspiration is a genuine new frontier. It is a growing body of techniques for making materials with novel and startling properties: surfaces such as paint and glass that clean themselves, fabrics that exhibit shimmering colour despite having no coloured pigments, fibres tougher (weight for weight) than nylon or steel based on spider silk, dry adhesives based on the microstructure of the gecko's foot.

It is not just a new frontier because these properties are startling but because they have something in common. The mechanisms of most of these effects are caused by physical structures of a certain size: from one billionth of a metre up to one millionth of a metre (fig. 1.1). This is the nanoregion and the structures nature builds at this level we can call nature's nanostructures. Until recently, the nanorealm remained relatively inaccessible to science and this may seem strange since scientists are able to manipulate subatomic particles millions of times smaller. And chemistry, a precise science with a growing inventory of more than 24 million discrete substances, operates at the size range just *below* the nanoscale.

The key to this paradox is that there is a huge gap between what we can *infer* about the size of atoms and molecules (and their even smaller constituents – protons, electrons and the like) by elegantly indirect experiments in chemistry and physics, and what we can *see* with the aid of a microscope. The ability of microscopes to magnify the smallest features has improved immensely since their invention in the late 17th century but there is a limit that is set by the properties of light itself.

When light hits objects patterned at just below one thousandth of a millimetre (1 micrometre or 1,000 nanometres) strange things begin to happen to it. This is because light itself is patterned on the same dimension. Light is a wave motion, with the peaks of the waves repeating at just below the 1 micrometre mark. When the waves meet patterns of a similar size, they bounce off in ways that blur the picture. This is known as interference and in itself it plays an important role in bio-inspiration (*see* Chapter 5).

As far as microscopy goes, though, this is simply a nuisance. With the light microscope we can see living cells and some of their contents – bacteria, spermatozoa, etc – but not the complicated large molecules that make up these structures.

Microscopy and chemistry began at more or less the same time in the late 17th century and closing the gap between them has been a long and tortuous business. At first, chemistry had nothing to do with size. The initial job was to identify which substances could not be broken down into anything simpler – these are the elements such as hydrogen, carbon, oxygen, nitrogen, sulphur. It was a matter of speculation as to what was the smallest possible part of an element. The best theory going at the time was the Atomic Theory that suggested that elements were composed of millions of identical tiny billiard-ball-like particles. For centuries, this was purely a theory. No one knew how large atoms were or if they really existed at all.

But, in the late 19th century, thanks to work on the pressure of gases,* it became possible to estimate the size of these 'atoms' (by now most scientists accepted that they existed). The first accurate figure for the size of individual atoms was made in 1908. Atoms are very small – in fact they are just off the nanoscale. A typical small atom such as carbon is about 0.3 nm (nanometre) in diameter.

So, if atoms were less than 1 nm in size and the smallest object you could see with a microscope was 1,000 nm, what existed in this Blind Zone? To try to understand how much we were missing, imagine being able to see objects, say, up to 1 cm but nothing more until you get to 10 m. Most of what we make and live with lies within this range (micro-electronics excepted). The equivalent for nature is the region ten million times smaller – and this zone was inaccessible to us.

Peering into this realm in the early 1960s, we were as blind as the moles in a fable by the Czech immunologist and poet Miroslav Holub: his poem 'Brief reflection on cats growing in trees' imagines the moles trying to make sense of the world. Lookouts emerged at different times of day to report on the way things were above ground. The first scout saw a bird on a tree: 'birds grow on trees', he reported; the second found mewing cats in the branches: 'cats grow on trees, not birds'. The conflict worried one of the elders, so up he went:

> By then it was night and all was pitch-black.
>
> Both schools are mistaken, the venerable mole declared.
>   Birds and cats are optical illusions produced
>   by the refraction of light. In fact, things above
>
> Were the same as below, only the clay was less dense and
>   the upper roots of the trees were whispering something,
>   but only a little.

* The pressure of a gas is caused by the billions of its molecules hitting any object it is in contact with. It is possible to calculate the size and speed of the molecules from the pressure, temperature and density of the gas. Molecules are compound atoms – an oxygen molecule, for example, consists of two strongly linked oxygen atoms.

'Things above were the same as things below', or vice versa in our case. We had only our knowledge of chemistry at the bottom and the world of visible objects at the top to guide us. When we look around we can see only such objects as can be seen with eyes like ours. We make use of materials that we can grasp and manipulate to make objects on a scale that suits creatures around 1.5–1.8 m tall. We may not like to think of ourselves as being as cramped in our perception as the moles, but on the scale of the universe, from quarks to galaxies, we are. In the scale of things, we are trillions of times larger than the smallest things known, evanescent subatomic particles, and trillions of times smaller than the largest cosmological objects known.

What exists in the Blind Zone are large molecules of complex non-random chemical composition that are assembled to make the working structures of the cell: pumps and engines and factories for making everything the cells need, including copies of themselves. The contents of the Blind Zone comprise nature's nanotechnology. And these are the nanomachines and structures we wish to harness for our own purposes.

But how could the gap be closed? How could we see nature's nanomachines at work? The answer was to nibble at the problem from both ends. As chemists gained in confidence throughout the 19th century, the chemical structures of some of the molecules used by living things began to be deduced: sugars, for instance, and the amino acids that are the ingredients of the fabulously complicated proteins. And as the 20th century progressed, the structures of larger and larger natural molecules were worked out.

Although the limitations of light microscopy were unbridgeable, even in theory, new techniques of investigation became available. By far the most important new investigative technique in the mid-20th century was the use of X-rays; with a wavelength thousands of times smaller than that of light (*see* fig. 5.2, page 105), these allow us to penetrate deep into molecules such as proteins. When X-rays hit molecules they produce complex reflection patterns that mirror the actual structure of the molecules themselves. Strangely, this reflection of X-rays is exactly the same property that sets a limit to light microscopy. The result of an X-ray analysis is not a photograph in the conventional sense. When X-rays hit a crystalline substance they are scattered in a regular geometric fashion and the patterns produced give information about the position of the atoms in the crystal. So this is not a picture so much as the result of complex mathematical analysis of data.

And it was a combination of chemistry and X-ray analysis that led to the greatest biological breakthrough of the 20th century, the elucidation of the double helix of DNA. The chemistry of DNA had already shown that it was composed of certain known substances: sugars and four different bases, with these bases, intriguingly, seeming to be paired. In any DNA sample, from whatever source, there was always as much adenine as thymine and as much cytosine as guanine. With this knowledge, it was possible for Watson and Crick to interpret the X-ray picture and to deduce the double helical structure.

From the 1950s onwards, this technique – the combination of chemistry and X-ray analysis – allowed scientists to work out the structure of many significant biological molecules, especially proteins. However, X-ray techniques are limited by the fact that the specimen has to be a crystal, and many biological molecules cannot be crys-tallized. And also, we want to *see* the larger structures that the molecules make up.

In a sense, the beginning of a sustained interest in the nanorealm can be dated precisely, for it was on 29 December 1959 that Richard Feynman gave *that* talk. Feynman's was a rallying call and it was heeded first in solid-state physics, as the relentless development of ever smaller and more integrated electronic circuits began. Finally, the better microscope requested by Feynman did arrive and biologists were allowed a glimpse into the nanoworld. This was the scanning electron microscope (SEM), invented in 1965 by Cambridge Instruments after decades of pioneering work at Cambridge University. Since then, many more advanced electronic instruments, such as the atomic force microscope, have followed, and a battery of different techniques can be brought to bear on natural structures. Ron Fearing, fabricator of gecko tape and micro air vehicles at Berkeley, University of California, talks of the 'psychological barrier that was broken in the sixties with micro-machining, the atomic force microscope coming along. Before, people would have looked at these structures and said, "Oh, that's too small to know what's going on"."

The SEM was a big breakthrough and it has had huge consequences for bio-inspiration. The pictures revealed by the SEM look like engineering of an exquisite kind. The organs of minute insects and the parts of plants are revealed as wonderfully tooled artefacts. Bio-inspirationists constantly have to track back and forth between the nanorealm and the everyday scale of things. According to the Russian novelist and serious amateur lepidopterist Vladimir Nabokov in *Speak, Memory*, this is an intrinsically artistic activity:

> There is, it would seem, in the dimensional scale of the world a kind of delicate meeting place between imagination and knowledge, a point arrived at by diminishing large things and enlarging small ones, that is intrinsically artistic.

When the first pictures were seen, the question of how nature achieved these wonders of micro-engineering was completely off the agenda – scientists could only goggle at the structures. But now we know a lot more about how nature creates such shapes. *The Gecko's Foot* is the story of how we are closing in on this last frontier of natural exploration.

The nanoworld is like a complex jigsaw puzzle in three dimensions. We try to piece it together by viewing it with different magnifications and techniques. Behind the picture we can see with the unaided eye, there is another picture we have to zoom in on with the light microscope; behind that is a more detailed picture that we need the electron microscope to see; beyond that is the picture revealed by X-rays; and there are new types of microscope, such as the atomic tunnelling microscope, that all add information to the puzzle. To add to this, our knowledge of chemistry also sheds light on the three-dimensional structure. By combining all the information, we come to a picture that begins to approach completeness.

In retrospect, it seems curious that we have been ignorant for so long about *how nature makes stuff*. While we are pretty good at making intricate structures ourselves, when it comes to the miracles of the human body our role in the construction process is crude and lumbering. Anne Stevenson's poem 'The Spirit is too Blunt an Instrument' makes this point:

> The spirit is too blunt an instrument
> to have made this baby.
> Nothing so unskilful as human passions
> could have managed the intricate
> exacting particulars …
>
> Observe the distinct eyelashes and sharp crescent
> fingernails, the shell-like complexity
> of the ear, with its firm involutions
> concentric in miniature to minute
> ossicles. Imagine the
> infinitesimal capillaries, the flawless connections
> of the lungs, the invisible neural filaments …

So, if not the spirit, what is nature's organizing principle? How does nature create intricate structures? There is still much to learn and our own attempts at mimicking these processes are fumbling, but we are now on the trail.

To understand why the realm of bio-inspiration is such a *terra incognita*, something really new under the sun, we need to look at the two great currents of 20th-century science. So powerful were these two prongs of attack that many people were dazzled into thinking that they revealed all we needed to know about the material world. These sciences were nuclear physics and molecular biology. Both ignored the multiplicity of the natural world – the several million species of living creatures (some estimates go as high as 30 million or more), all with different shapes, sizes, habits and curious adaptations; the more than 24 million known chemical combinations of the 92 natural elements; the architecture of matter in the honeycombs of the beehive, the fantastic filigree forms of the radiolarians of the ocean, and the interlocking spirals of a sunflower head. These were cast aside in the search for the ultimate, universal components and principles of matter (physics) and the chemical unit and mechanism of genetic inheritance in biology.

The idea behind these quests was that if successful, they would somehow explain everything else. And, of course, they *were* successful. Nuclear physics uncovered the unexpected power of nuclear forces and molecular biology determined the mechanism of inheritance: a precise sequence that has a chemical form (the DNA molecule) but which functions as a code for the synthesis of proteins, nature's prime functional substances.

But, dramatically brilliant as these sciences were, they left enormous gaps. They did not begin to explain complex forms of nature, nor did they determine the composition of these forms. What the physics and biology obscured was the fact that to create functioning organs, the fundamental building blocks of atoms and molecules have to be synthesized into large structures whose properties cannot really be explained by a knowledge of which molecules compose them. The biologist Helen Ghiradella wrote in 1991, just before the bio-inspired explosion:

> Many of us working in biological fields have perhaps unconsciously assumed that small things must be simple, at least more accessible to human understanding than those on a human scale. This may not be the case, and indeed, the further we investigate the more complexity we seem to find.

When, as a schoolboy in the early 1960s, I became fascinated by chemistry, what I wanted to know was: What are familiar objects made of? How is a tiny insect engineered from biological materials? What is the chemical structure of wood? What, in chemical terms, is a spider's web? In *The Periodic Table*, Primo Levi beautifully expressed this chemist's lust to know the fabric of the world:

> Everything around us was a mystery pressing to be revealed: the old wood of the benches, the sun's sphere beyond the window panes and the roofs, the vain flight of the pappus down in the June air. Would all the philosophers and all the armies of the world be able to construct this little fly?

But, at the time, chemistry had no answers to these questions.

Whenever such structures and substances were mentioned in textbooks, the explanations petered out in sentences such as: 'The hardness of the insect skeleton is due to the chitin being impregnated with another substance, called sclerotin or cuticulin; but not much is known about it chemically.' There were some successes in getting close to nature. Nylon, for instance, invented in 1937, imitated the chemical bond of natural protein fibres, but natural proteins such as wool, silk and spider silk were known to be much more complex than nylon. While the nylon molecule has the same chemical unit, linked nose to tail thousands of times, natural silks have different amino acid units, linked nose to tail in a complex non-random pattern. Despite a concerted effort over the last 20 years to determine the structure of, and replicate, spider silk, it is still not fully understood.

Although science has been successful in uncovering things not directly known to our senses, the mindset required to solve the problems of nuclear physics and genetic inheritance tends to be impatient of such questions as: What lies between the molecular realm and the objects we can see? The great early 20th-century nuclear physicist Sir Ernest Rutherford notoriously used to say that 'all science is either physics or stamp collecting'. But our new science has arisen largely from the very stamp collecting Rutherford despised – descriptive biology, investigations of the habits of strange creatures, comparative studies of the microstructures of leaves.

For a Rutherford, these meanders off the central pathway were expected to be explained fully by the fundamental laws of physics. And when his kind of particle physics was at the forefront in the mid-20th century, there were no techniques available to investigate larger-scale phenomena.

The atoms of physics and chemistry are very small (about one ten billionth of a metre in diameter) and until 1971 this was far too small for any kind of microscope to see. Their size and properties were inferred from experiments on much larger quantities than single atoms: 1 gram of carbon contains about fifty thousand million million million atoms (usually written $5x10^{22}$ to avoid the cumbersomeness of the expression). It was the triumph of chemistry that it was not necessary to see these tiny atoms in order to synthesize millions of new compounds whose precise structure is known.

This chemistry of inference, working in the dark, so to speak, was the chemistry I was taught at school in the 1960s – experiments were carried out with simple substances, stuff you could grasp and whose properties were clear. I might bubble carbon dioxide through limewater, say; the result was a white precipitate of solid matter. I could filter and dry this and the result would be calcium carbonate, the chemical that chalk and limestone are made of. Running parallel to this palpable experience, the books would give you an equation for their reaction, in this case:

$$Ca(OH)_2 + CO_2 = CaCO_3 + H_2O$$

Like mathematical equations, these equations always balance because they represent the reactions between individual atoms and molecules, and nothing is ever lost in a chemical reaction. There is one calcium atom, four oxygens, two hydrogens, and one carbon on both sides of the equation. What happened in my test tube was this reaction, between individual calcium hydroxide and carbon dioxide molecules. And it was happening billions of times over to make enough of this substance for it to be visible to my eyes.

I chose this reaction as an example because the simple minerals of school chemistry, such as calcium carbonate and silicon dioxide, turn out to be capable of forming structures of architectural complexity in living systems, many of which are to be found in the deep oceans. The extreme conditions to be found there – intense pressure and little light, the dispersed nature of prey, the single medium of water – have inspired some ingenious devices. The romance of the oceans is epitomized by the Venus flower basket, a sea sponge and a baroque extravaganza of mineral basketwork so ornate that Joanna Aizenberg, the biomineralization expert at Bell Labs, who is studying it for its fibre-optical properties, cannot yet see how such a structure can grow from an egg. A new frontier indeed! To its beauty and mystery have now been added the fact that it possesses in the long hairs that surround the base of its latticework some brilliantly effective fibre-optic filaments. These, in human engineering terms, are the conduits used for high-capacity telephone and internet lines. The Venus flower basket has evolved these structures to manipulate what little light there is on the sea floor (at least we think it has – as with much else about the creature, biologists are not entirely sure).

Then there are the brittlestars, with primitive eyes that focus light through exquisitely engineered lenses made from single crystals of calcium carbonate (*see* Chapter 5). In these creatures, the crystallization of calcium carbonate is directed by proteins and this is one of the prime routes being explored in bio-inspiration: to direct the formation of engineered structures of minerals such as calcium carbonate and silica, using proteins, as nature does.

But simple chemistry was inadequate to explain how proteins organize minerals to produce these complex forms. Proteins are, unlike calcium carbonate, very large molecules. The molecular weight of $CaCo_3$ is 100 D (D stands for 'Dalton' and is a measure of the relative mass of atoms and molecules, hydrogen being 1 Dalton) but a protein can contain thousands of different amino acid building blocks in one molecule, and the molecular weight might be 300,000 D.

Although attempts to derive engineering solutions from natural mechanisms have only begun to be made in the last 15 years, earlier biologists came close to guessing their potential. Sir Alistair Hardy, in *The Open Sea* (1956), repeatedly marvelled at natural mechanisms as feats of engineering. This is Hardy on the stinging hairs found on many jellyfish:

> It is not a living thing; it is a dead structure, an elaborate tool made ready for work – and made to perfection – by the semi-fluid living substance of the cell. Here is something to wonder at, for it looks as if it were designed.

Behind this you feel the lurking suspicion that *we* ought to be able to design such a structure. In this case, we haven't yet done so but the action of biological springs like the jellyfish's sting is definitely on the bio-inspired agenda. Hardy has the true spirit of bio-inspiration before its time. My second-hand copy of *The Open Sea: The World of Plankton* (*The Open Sea* is in two volumes, one on the world of plankton, the other on fishes) has an interesting history. It is stamped inside: 'MoD Library Services: withdrawn from stock.' These days, the Ministry of Defence is a principal funder of work in bio-inspiration. I hope they have bought a new copy.

Bio-inspiration has an appeal denied to other cutting-edge sciences. Firstly, it involves some attractive creatures, adding an extra dimension to the allure of butterflies, geckos, lotus plants and the like. Then there is the utility of the products – this is a technology, not science for science's sake. Bio-inspired solutions are often comprehensible in a way that much science is not: they involve structures whose functions are clear, even if they need a microscope to see them. Finally, some subjects of bio-inspiration are amenable to kitchen-table experimentation, as this book will demonstrate.

Inevitably with a new subject, there is some uncertainty about the boundaries of bio-inspiration. Scientists working in bio-inspiration generally fall into one of two camps: biomechanics or materials science. Biomechanics is concerned with large-scale mechanisms, such as how insects fly, materials science with fine-scale structure and chemical composition. It is worth remembering that, historically, these two disciplines come from very different traditions. The materials scientists prefer the term 'biomimetics' for this new subject but the biomechanics don't like this because it suggests to them a slavish copying of nature (*mimesis* = 'copying'). When he lectures, Professor Bob Full, the ebullient master of animal locomotion at Berkeley, University of California, even has a slide with a big red slash through the word: 'Biomimetics? No,' he says, 'Bio-inspiration is the way to do it.' In an important sense Full is right. Scientists try to unravel nature's mechanisms, but technologists use whatever will work. Bio-inspired technical products will almost certainly not mimic the actual materials used by nature. The self-cleaning Lotus-Effect® (*see* Chapter 2) is the most advanced of these techniques in terms of coming to market, with several products available, but it does not use the actual substances found in lotus leaves.

It is worth thinking about how nature and the human engineer went about producing their structures before we reached this point of rapprochement at which engineers are eager to learn from nature. Design in nature and in engineering are achieved by totally opposite methods. The human engineer can start from scratch, designing on paper something never seen before and then assembling the parts until it is all connected up and ready to go. For example, for birds to reach their present sublime level of design, it has taken millions of years of evolution. In the 1940s, aeroplanes made the abrupt jump of moving to jet engines from piston engines that drive propellers. The jet engine was perfected by Frank Whittle between its invention in 1928 and the first flight in 1941. If nature had wanted to evolve towards something similar there would have been an intermediate creature that could still fly by the old method while the new one was developing.

Bob Full makes the point like this: 'If I told you to take my '84 Toyota and make it the fastest car possible using any material that you have, you could make a pretty fast car if you could replace 20 things. But you can't throw away the whole genome and start from scratch. That's a pretty heavy compromise.'

Put like this, it would seem that the human engineer holds all the aces. If, as a designer, nature is hobbled in this way, surely the human engineer ought to win hands down? But, despite her apparent constraints, nature has still produced devices for which engineers would give their eyeteeth. With regard to flight, for example, human aviation is impressive but in terms of manoeuvrability, the fly leaves a modern jet fighter standing, being able to turn a right angle at speed in only one twentieth of a second.

We are fortunate that we can have it both ways, using nature when it has developed structures we can adapt, while at the same time retaining the engineer's radical risk-taking advantage over evolution's necessarily conservative processes.

There are times when it seems that bio-inspiration should be called 'technomimetics': only too often physicists, engineers or chemists invent something; biologists then discover that nature has already invented it (often hundreds of millions of years before), but the phenomenon itself was not known until discovered by the technologists! Obvious examples are echo-location in bats and sonar in whales and dolphins: before ultrasound was invented scientists could have dissected bats for eternity and still not understood their echo-location mechanism.

The most dramatic recent example has been the photonic crystal: a nanostructured crystal that will enable light to be guided at fantastic speeds through the crystal to create pathways in which information can be stored and manipulated. The photonic crystal was predicted as a theoretical possibility by physicists in 1987, first created technically in 1991, and discovered in butterflies and marine creatures in the late 1990s. In other areas, biological discovery has led to technical invention in the true spirit of bio-inspiration. In fact, in the case of the Lotus-Effect, once the biological effect was established – that some plant leaves have a micro-structure that produces highly developed water repellency and self-cleaning – it was realized that physicists had produced a general theory to account for this 50 years earlier, but its importance had not been recognized. Now, super-water repellency is a respectable subject in many physics and materi-als science laboratories where you won't find a leaf of any kind.

Bio-inspiration is not a narrow discipline. Origami was once thought merely to be an amusing game, nothing to do with science. Then mathematicians realized that it could be interesting to them, as a branch of topology: the maths of shapes. Origami is used by nature because some structures such as leaves and wings need to be folded. Now whenever human engineers want to deploy structures (erect something that is usually kept folded), they look at the ways nature uses origami.

Although Primo Levi, the great Italian writer and chemist who died in 1987, did not live long enough to see the birth of bio-inspiration, he did have an abiding interest in the natural world. He was especially fascinated by insects and in his essays (*Other People's Trades*) he said of beetles:

> These small flying fortresses, these portentous little machines, whose instincts were programmed one hundred million years ago, have nothing at all to do with us, they represent a totally different solution to the survival problem.

But beetles, like every other major group in the natural world, *do* have something to offer us. The flashing light of the firefly (a beetle despite the name) is caused by a chemical reaction that produces almost no heat and this has been mimicked to produce biomedical diagnostic tests. The Oxford zoologist Andrew Parker has discovered a desert beetle that has a novel way of capturing the sparse water that comes its way and this too will have technical applications.

And what of the bombardier beetle, a creature that seems to have anticipated many of the principles of human rocketry? It has a powerful defence mechanism that involves directing a hot irritant spray in the direction of an attacker. The chemical propellant for the spray turns out to be hydrogen peroxide, a well-known human rocket fuel. The peroxide is mixed with hydroquinones in a 'reaction chamber'; the reaction is hot (80°C) and the gases produced result in an explosive exhaust. The reaction chamber can be swivelled like a rocket motor to point towards the attacker. The whole business sounds far more like human technology than a natural creature. The more we know about beetles the more they seem to be little compendia of bio-inspirational properties.

Bio-inspiration can work across the whole size spectrum but there is no doubt that most of the work presently being carried out is in the former Blind Zone, the nanoregion. The idea that the properties of things as experienced by us derive from tiny structures goes back a very long way: back to the 5th-century-BC Greek philosophers Democritus and Leucippus who proposed the atomic theory of matter. Their ideas are known to us through the exuberant epic poem *De Rerum Natura* by the Roman poet Lucretius (*c.* 100–*c.* 55 BC). Lucretius would have loved bio-inspiration. He tried to answer fundamental questions: What is the world made of? Can matter be created or destroyed? Are conscious beings made of conscious stuff? How does life renew itself? And despite three centuries of modern science that would have astonished Lucretius, many everyday things remained unexplained until recently. As he makes clear in *De Rerum Natura*, he was aware of the mystery of nature's tiny functioning organs:

> How small can anything be? We know of creatures
> So tiny they would seem to disappear
> If they were less than half their present size.
> How big do you suppose their livers are?
> Their hearts? The pupils of their eyes? Their toes?
> Pretty minute you must admit.

Lucretius believed that the underlying particles of the material world could not have the same properties that appear to our eyes. They *had* to be colourless, odourless and tasteless (and lacking consciousness). It was a subtle idea; you might think that if you kept chopping something up until it was very small it would be the same all the way through – just smaller – but it was Lucretius's intuition (I refer to this as the Lucretian Leap) that, at the smallest scale, things just had to be different. In this sense, Lucretius and his forebears were the first nanotechnologists, although the subject had only a notional existence in their imaginations.

De Rerum Natura gives us, in a language all can understand, a passionate explanation of the way things are. Indeed, it is unfortunate that modern science has not proved amenable to the Lucretian treatment. But bio-inspiration is remarkably Lucretian in spirit. It answers simple bold questions about aspects of nature: Why is the lotus leaf always clean? How does the gecko walk upside down? How can a spider's web be stronger than steel? How can a fly of little brain be more manoeuvrable than a Eurofighter?

The answers to these questions are also Lucretian. Lucretius constantly argues that the *causes* of the effects that we see are different in kind to the effects. This could almost be the first law of bio-inspiration: the tidiest surfaces are the roughest at the nanolevel; the structures that cause the colour of the peacock's tail are *not* coloured; the hairs on the feet of the gecko are *not* sticky.

Take silk, a byword for slinkiness. But what is the gorgeous crackle it makes when you rub it against itself (known as 'scrooping') and what causes the colour changes when dyed silks are viewed from different angles? Early synthetic silks did not have these properties because the fibres were smooth and of rounded section. But under the SEM a natural silk fibre will be observed to have micro-structured rough edges – not at all what might be expected from the feel of it. When, in the 1980s, Japanese textile manufacturers realized this, at last they were able to make close synthetic copies of natural silks: they called them *Shin-Gosen* ('New Feel').

Lucretius was not the only poet whose imagination was caught by these natural phenomena. In his poem 'Greatness in Little', the 17th-century English poet Richard Leigh was intrigued by tiny things, long before such a fascination could be satisfied. At one point he bursts into praise of minuteness itself:

> Ah, happy littleness! That art thus blest,
> That greatest glories aspire to seem least.
>
> Even those installed in a higher sphere,
> The higher they are raised, the less appear …

Bio-inspiration usually works at the nanolevel, but that does not make it synonymous with nanotechnology. Most nanotechnology is not bio-inspired; it is the province of materials technology, comprising things like smaller electronic components and nanoparticles in cosmetics and systems for delivering targeted doses of drugs. Another name for bio-inspiration makes clear the distinction: bio-inspiration is 'nature's nanotechnology'.

What is especially interesting about nanotechnology and bio-inspiration is the existence of hybrid technologies – systems in which one part comes from technical nanotechnology and the other part from natural mechanisms. Our minds are attuned to an 'animal, vegetable, mineral' classification system and we generally assume that anything will belong to just one of these categories. One of the most dramatic discoveries of bio-inspiration is that technical components can be incorporated into natural systems.

Although cells never meet silicon chips or other electronics materials (which have only existed for 30–40 years) in nature, natural proteins can stick to silicon and other electronics materials and in doing so create structures on a much finer scale than would be possible for the technical materials alone.

Engineers make things by heating, beating and hacking them into shape; chemists make things by cooking up the ingredients; nature makes things through the DNA in the genes. The plan of every creature that exists is at some stage just a coded blueprint strung out along the double helix of DNA. Nature's way is by far the most subtle, accurate and fine scaled. It can be seen as a combination of engineering and chemistry. DNA is a chemical but it also has architecture – the double helix – and the substances DNA makes –

proteins – also have architecture. They are both chemicals and pieces of nanoengineering.

To exploit DNA's design potential, bio-inspirationists use hybrid techniques of genetic engineering and silicon-chip fabrication. This may trouble some people – but nature does not recognize the division between organic and inorganic: gorgeous *inorganic* mineral shell structures are produced under *organic* control. The rigid organic/inorganic divide is a product of the human mind, more specifically that of chemists who have put the labels 'Organic' and 'Inorganic' over the doors of departments which used to have very little to do with each other.

When I put it to Mehmet Sarikaya, the passionate advocate of this new hybrid technology, that some would see a Frankenstein element in it, he said: 'There will be more good happening than bad because human beings are fundamentally good people. What's at the back of this scientist's mind is: Can I have an impact on the early detection of cancer? Can I have an impact on assembling nanofibres for new nano-molecular devices? That's what we have in mind, that's why we work.'

Although bio-inspiration is still largely unfamiliar to a wide public, nanotechnology has already attained a certain notoriety. The idea is abroad that there is something inherently dangerous in the nanorealm. Michael Crichton's bestseller *Prey* (2002) imagines self-replicating nanorobots escaping from human control, learning rapidly and becoming ruthless predators. What lies behind this fantasy and does it have any credibility?

We have always lived in a nanoworld – our bodies and those of all living things are composed of biological nanomachines, and the dust in the air, pollen grains, smoke from all forms of combustion, contain nanoparticles – but like M. Jourdain in Molière's *Le Bourgeois Gentilhomme*, who was astonished to discover that he had been speaking prose all his life, we have only just woken up to the fact. And this has caused panic in some quarters. The idea behind *Prey* came from Eric Drexler's *Engines of Creation* (1986), which first put nanotechnology into the public arena. Drexler suggested that nano-technology would spawn self-replicating systems that might get out of control, thus swamping the world with a 'grey goo' of synthetic nanomaterial.

The idea of 'grey goo' took on a life of its own; it was resurrected in 2003 by Prince Charles in a speech warning of the dangers of nanotechnology. But, in 2004, Drexler set the record straight, in an article co-written with Chris Phoenix of the Centre for Responsible Nanotechnology, saying:

> Nanotechnology-based fabrication systems can be thoroughly non-biological and safe: such systems need have no ability to move about, use natural resources, or undergo incremental mutation. Moreover, self-replication is unnecessary; the development and use of highly productive systems of nanomachinery (nanofactories) need not involve the construction of autonomous self-replicating nano-machines.

Of course, the nanotechniques of bio-inspiration *are* biological but, when you look at what these techniques are, you will see that there is no way that their products could reproduce themselves and get out of control.

Even if nanotechnology is not going to swamp the world, many people remain concerned about some aspects of it. In 2004, the Royal Society and the Royal Academy of Engineering published a report on its benefits and possible dangers. The report stressed that while it would be wise to be wary of ingesting nano*particles* and releasing them to the environment without tests to ascertain what effects these substances can have, nano*structures* are a different matter. Nano*particles* are potentially dangerous on two counts: being so small they can enter cells by routes forbidden to larger particles and because they have such a large surface area relative to their volume their chemical and electrical properties are enhanced, which raises the possibility that they could trigger damaging reactions within the cell. There is no reason to fear *solid* objects structured at the nanolevel: the world is full of solid nanostructures. All living things, including us, are necessarily nanostructured – made from atoms which have to be assembled into nanostructures before they can make up anything large enough to be seen.

The possibility of a hysterical reaction to things nano really came home to me when I visited the glassmakers Pilkington in St Helens,

Merseyside, to discuss Activ™, their new self-cleaning glass, described in Chapter 2. Activ glass has a very thin coating on the surface that gives it self-cleaning properties. This coating is less than 20 nm thick. Kevin Sanderson, one of Activ's inventors, told me that they had received worried telephone calls asking, 'Are these nanoparticles on my Activ glass window going to fly off the surface and do me harm?' In fact, the nanolayer is bonded very strongly to the glass underneath, it is harder than glass and will last as long as the window does.

As for nanoparticles, there have always been and always will be nanoparticles in the environment: they are called dust. All forms of combustion produce huge clouds of them. The air in the London Underground is full of nanoparticles and some of them may even be carbon nanotubes, the most famous nanoparticle, created by the action of electric sparks from the live rails. It seems likely that the concern about nanoparticles being added to sunscreens and cosmetics will lead to new research on our *total* exposure to nanoparticles – from car exhausts and the Underground, to bonfires and barbecues.

Every new technology creates fear and resistance, but as far as it is known at all, bio-inspiration has had a good press to date. It has an eco-friendly feel to it, unlike the more hard-edged nanotechnology; but once its connection to nanotechnology becomes known – that bio-inspiration is mostly nanoscale technology – it will be damned by some through association. So it is important to stress that there is nothing wrong with nanotechnology *per se.*

But if nanotechnology induces fear in some people, in science it is also a buzzword: play the nanocard and you unlock the funders' purse strings. As a result, a lot of people have suddenly discovered that, in reality, they are doing nanotechnology. Physicist Andre Geim tells of engineers 'who never make anything smaller than 1 metre in diameter and now they're doing nanotechnology because they can *position* their things to within 1 nanometre accuracy!' As Geim says, 'Adam and Eve were nanotechnologists, they created everything from sperm, from DNA!'

Bio-inspiration arrives at a time when there is organicism in the air, especially with regard to architecture and design. The theme of

Expo 2005, held in Aichi, Japan, from March to September 2005, was 'Nature's Wisdom'. Organicism is abroad in both the *Zeitgeist* of general culture and in materials science and there are connections between large-scale bio-inspired architecture and bio-inspired materials. Many architects want to design smart buildings and to use the new bio-inspired materials. The first really commercial application of bio-inspiration is in paint for the exteriors of buildings, using the Lotus-Effect, closely followed by Pilkington's self-cleaning Activ™ glass. And if other bio-inspired materials are not yet ready, there is no law against including organic curves in the shape of a building.

In September 2003, the *Zoomorphic* exhibition at the Victoria and Albert Museum recognized this new tendency in architecture, with structures based on many creatures, from sea sponges to dinosaurs. The archetypal figure is the Spanish engineer and architect Santiago Calatrava, creator of the Athens Olympic Stadium. Calatrava's buildings and bridges exhibit creaturely gestures rather than mimicking specific creatures: there are moth-like antennae, forest canopy train-shed roofs, reptilian snouts, a whale's tail (or bird of prey's wings). After the turbulent history of architectural styles since the early 20th-century modernist revolution, organic architecture seems an attractive option. It uses the same materials as hi-tech architecture, and both organic and hi-tech architectures have their roots in geometry. Indeed, the key to all bio-inspiration is that nature and human artefacts are acted upon by the same forces and they occupy the same three-dimensional world. And this is why similar solutions are possible in each.

Alongside the architecture, in cars such as the Vauxhall Tigra, Ford Ka, Volkswagen New Beetle and the latest Nissan Micra, recent car design has also shown itself leaning towards organicism. The idea behind these cars is to be *expressive*: they sit unusually on the road, with the tail up, and the headlights styled as eyes, giving the impression of a face. These are cars whose moods you can read. In the case of the Vauxhall Tigra, the first of the breed, there is a resemblance to the warning display of an eyed hawkmoth – which is appropriate, because the hawkmoth displays large eye patterns on its wings, trying to look like a much larger and fiercer creature;

similarly, the Tigra is a tame little Corsa dressed up to be racy. Whether or not there is a *functional* reason for such large-scale organic structures (and often there is not) they belong to the new worldview that bio-inspiration has ushered in.

While writing this book, I have found myself watching insects in the garden far more closely. In fact, I wonder if I ever really noticed them before, other than on the increasingly rare occasions that a butterfly flew in. A sudden flurry in the corner of my eye and a garden spider is binding an already unrecognizable insect. Hoverflies punctuate the air around the *Coreopsis*. Two cabbage whites lurch across the garden in a mating dance. I realize that one of the reasons I used to be impervious to this micro-choreography is that it all seemed so impenetrable. How on earth did they do it? But, increasingly, we know, or if not, we know *how* we are going to know in a few years' time. Welcome to an Aladdin's cave of bio-inspired materials and devices.

CHAPTER TWO

The Great Sacred Lotus Cleans Up

Though buried deep
In the slime of the pool,
Unstained and untouched
You come forth to the world
Glorious in beauty,
Pure and serene:

Yet in your innocence
Oft you deceive us
Transforming the dew
On your life-giving leaves
Into sparkling gems!

GONNOSKÉ KOMAI, 'To the Lotus-Bloom'

'Nooks and crannies harbour dirt,' we have always been told: a piece of folk wisdom scientists would not have bothered to dispute until some 15 years ago. But the self-cleaning powers of the sacred lotus plant – recognized and sanctified thousands of years ago in the East – have turned this on its head. The lotus's secret is that its surface is *rough* at the micro- and nanolevels. It is almost embarrassing that such an elemental discovery should have waited so long to be made, but it has opened up for human use a new field of self-cleaning surfaces, utilizing the Lotus-Effect®.

Water skitters off a lotus leaf like drops of mercury – it doesn't spread and the globules it forms are highly spherical. So water doesn't last long on a lotus leaf. As for dirt, it seems to have a greater affinity for water than for the leaf so when it rains it is simply washed off.

There is a school of thought that science has still to rediscover the greater wisdom of the Ancients. In the case of the lotus, they are right. In ancient Eastern cultures, the lotus's immaculate emergence from muddy water was more than noticed: the plant became a symbol of the triumph of enlightenment over the dross of earthly life. So deeply does the lotus pervade Indian, Chinese and Japanese consciousness that the name is a byword for, and a guarantor of, purity. The most famous Buddhist chant, *Om mani padme hum,* translates as 'Behold! The jewel in the lotus', and the classic Buddhist texts are known collectively as the Threefold Lotus Sutra. The quest for spiritual cleanliness that runs through Buddhism derives from the lotus's example, so much so that images of cleaning recur in the texts:

> The Law is like water that washes off dirt. As a well, a pond, a stream, a river, a valley stream, a ditch, or a great sea, each alike effectively washes off all kinds of dirt, so the law-water effectively washes off the dirt of all delusions of living beings.
>
> *Innumerable Meaning Sutra*

While researching this book, I experienced my own lotus epiphany. I had flown from San Francisco to Seattle, and was en route from the airport to the University of Washington campus. It was a long day, my trip was almost at an end and I was tired and anxious. I had to change buses in the middle of Seattle's downtown subway system. I emerged in the middle of Chinatown and walked into the nearest café for a bite to eat. In the middle of the counter, staring up at me, were lotus cakes. I ate one – it tasted rather like chestnut – and a Proustian madeleine feeling came over me, although this was not for the recollection of time past but a kind of blessing on the future of my enterprise. I had risen from the underworld of the subway system, in which the route to enlightenment – Washington University campus – was temporarily lost. The notion of sweetness arising from dross is such a powerful one that once you know of the lotus you cannot help but refer to it: hence its omnipresence in East and South Asian cultures.

In the West, appreciation of the lotus is more aesthetic than spiritual: 'No more stately plant adorns our gardens than lotuses,' is a typical statement from an early 20th-century horticultural book on the water lilies.* Concerning the flowers, the book goes on: 'These great blossoms are among the noblest products of the vegetable world. They fairly glow in the morning sunlight.' With flowers 20–30 cm across, some of the leaves sit on the water, as water lily leaves do, and some stand 1 m from the surface. The water that collects on them is tossed back into the lake by the wind. In size they are dwarfed by the largest water lily, the *Victoria regia* from the Amazon, which was first brought to flower in England by Joseph Paxton in 1849, but the grace conferred by the lotus's exceptional purity more than compensates for that. (Incidentally, *Victoria regia* also has a role in the development of bio-inspiration; Paxton, as the engineer of the Crystal Palace in 1851, was much influenced by its structure; *see* Chapter 9.)

I was not sure whether I had ever seen a lotus before I became interested in the Lotus-Effect: water lilies of course, but had some of these been lotuses? I went to the Botanic Gardens at Kew, London, to find out for myself. Lotus plants die down every year and in cultivation are replanted from the runners that spread from the rhizomes rooted in the mud. At Kew in April they had plants of a variety of the American lotus, 'Perry's Giant Sunburst', growing in tanks next to water lilies. Although it had a name redolent of out-of-town garden centres, nevertheless it was a real lotus: the leaves had

* In evolutionary terms, lotuses and water lilies are not related; the similarities between them are due to their parallel evolution, in which the strict requirements of an environmental niche have led to similar species evolving from different ancestors (a phenomenon vividly demonstrated by Australian marsupials: there is a marsupial equivalent for almost all of the common mammals, even though they have evolved completely independently). The lotus has a virtually identical American cousin and the fossil record shows that the lotus was once found in a broad band from Asia through Europe to America: the lotus became extinct in Europe during the Pleistocene ice ages (1.9 million to 10,000 years ago).

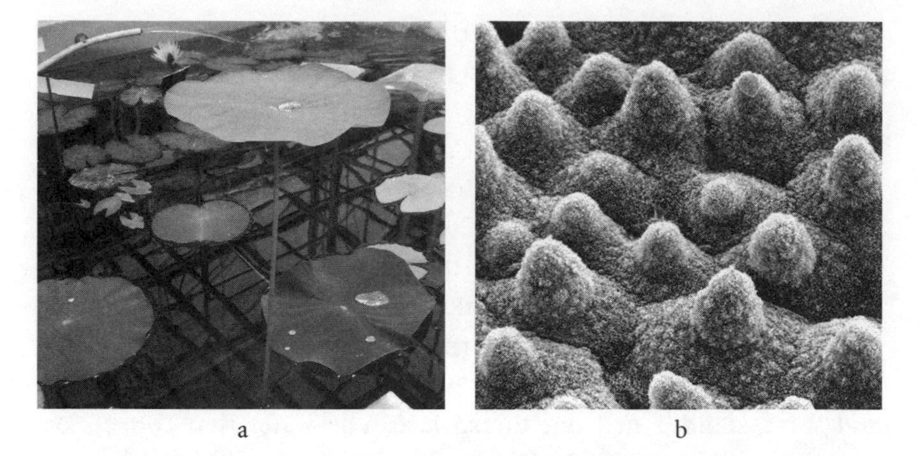

a b

Fig. 2.1 a) Leaves of the lotus 'Perry's Giant Sunburst' at Kew Gardens with 'quicksilver' water drops; b) the surface of the lotus leaf seen under the electron microscope.

that bluish bloom you see on some cabbage leaves. Dropping water on the lotus leaves was like dropping mercury on the table. The water drops gleamed with internal reflection and skittered around like quicksilver (fig. 2.1).

The Lotus-Effect's discoverer, Professor Wilhelm Barthlott, Director of the Nees-Institute for Biodiversity at Bonn, Germany, is unusual in pursuing parallel careers as a research botanist and as a patent-holding industrial inventor working closely with many industrial partners. 'Technology transfer' is a buzz phrase in universities these days, as governments try to kickstart economic growth by applying university expertise to the commercial world. The Lotus-Effect is a model of how it should be done.

Wilhelm Barthlott had no intention of becoming a technologist. He is a benign, avuncular and energetic man with bristling bottlebrush hair and a moustache that perhaps evoke some of plants he encounters. He has made a particular study of cacti and his interest in biodiversity stemmed from visits to Madagascar, where many of the plants are unique to the island. As often happens in life, Barthlott found the Lotus-Effect when he was looking for something else. Evolution was his obsession and in those days – before the emergence of molecular biology in the early 1960s – evolutionary relationships were studied purely by comparing the anatomy of creatures, especially their micro-anatomy: pollen grains for example. So Barthlott spent a lot of time at the microscope.

But then the scanning electron microscope (SEM) arrived that was to transform his work and would ultimately lead to his discovery of the Lotus-Effect. The SEM, which came onto the market in 1965, uses television-style scanning to produce richly contoured images with the appearance of 3-D.

With the SEM, a wonderland of fine structure, as detailed as any architect's fantasy, came into view. The surface of plants is a strange other-worldly terrain. The outer surface does not consist of living cells but a non-living shell, the cuticle, covered in layers of waxes of

---

Fig. 2.2 A gallery of bizarre structures on the surfaces of the leaves and seeds of plants: a) a star-shaped hair structure on the leaf of *Virola surinamensis*. It is surrounded by waxy bobbles although there is no wax on the star structure itself; b) these chimney-like waxy structures grow around the leaf pores of *Colletia cruciata*; c) pasta-like waxy crystals on the leaves of *Williamodendron quadrilocellatum*; d) interlocking cells on the seed coat of *Lychnis viscaria*.

varied composition. Sometimes the waxes are deposited on the surface in bizarre shapes (fig. 2.2). Through the microscope these structures often look more like animals than plants: *Virola surinamensis* seems to have miniature starfish nestling on a bed of waxy bobbles; the surface of *Colletia cruciata* resembles nothing so much as Anthony Gormley clay figurines, lolling about on the leaf; and *Williamodendron quadrilocellatum* has little piles of wax rings that could be a new form of pasta. Then there are miraculous architectural sweeps – the seed coat of *Lychnis viscaria* has plates that lock together like the tessellations of an Escher drawing. (Chapter 9 explores how structures like this have become important sources of inspiration for contemporary architects.) But most plants have bobbles like miniature topiary yew trees, with a frosting of waxy crystals on top.*

For a while, Barthlott was engrossed in the sheer beauty of these structures, but then something unexpected emerged. Specimens must be cleaned to be looked at in detail – at very high levels of magnification, contaminants can ruin the picture. But, in 1974, Barthlott realized that certain plants never seemed to need cleaning and that these, under the microscope, were always the ones with the roughest surfaces.

This was the beginning of a trail that was to take Barthlott far from his comparative studies of the structure of plants (although he is still highly productive in this field), into the world of technical production of a new invention. The full impact of the self-cleaning effect crept up on Barthlott over a long period: the early work, he says, was 'purely descriptive, without measurements'. He believed he had discovered something important in botany but 'it never occurred to me that it could be something new to physicists and materials scientists'.

* The strangest thing about the waxes was discovered by Barthlott's team in 2001, long after the Lotus-Effect had been recognized. If a plant surface is covered with a plastic film, the waxes can pass through the film and assume their customary shape by a process of self-assembly. It seems that in nature the waxes pass through the cuticle together with gases from the plant's breathing processes. As far as the plant is concerned, the plastic film is no different to the cuticle.

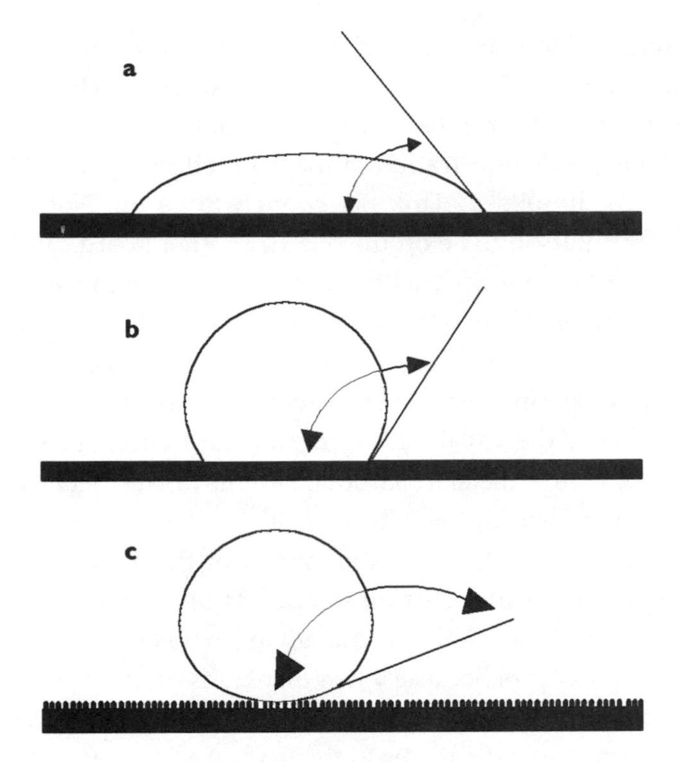

Fig. 2.3 Typical contact angles of water on: a) a water-loving (hydrophilic) surface – less than 30°; b) a water-repelling (hydrophobic) surface – greater than 90°; c) a Lotus-Effect® (super-hydrophobic) surface – greater than 150°.

So what is happening on the rough surfaces of those leaves? The self-cleaning effect depends on the relative 'wettability' of a leaf. Wettability is something we all recognize but scientifically it is something quite specific. On wettable surfaces, water drops are severely flattened and the contact angle that water makes with the surface of the leaf is very low (fig. 2.3). On a highly non-wetting surface, water forms near-spherical drops and the contact angle is very high – almost 180°.

When a surface has many tiny bumps, and these bumps are formed from a water-repellent substance, water drops 'sit' on top of the bumps, cushioned by the air in the space beneath them. The area of contact between the water and the surface is dramatically reduced by these bumps. The curious properties of an array of bumps in providing a cushion for an object sitting on them is demonstrated by the 'magic' illusion of the Fakir-on-the-Bed-of-Nails. The mystery of how the fakir can bear to lie on the bed of nails is no mystery at all.

In a standard demonstration of the 'fakir effect', about 1,000 nails are punched through a plank big enough to lie on. Not only is it possible for a person to lie on the board, another board can be piled on top to create a sandwich, a breeze block placed on the recumbent's chest, and the block smashed with a hammer. (The only danger to the victim – and to the block smasher – is flying debris: goggles must *always* be worn in this experiment.) The weight of the body distributed over the 1,000 nails does not exert enough force at the points to puncture the skin, although we intuitively feel that nails, however many there are, *must* be painful.

To translate from the large-scale world of the fakir down to the lotus surface: water drops sit on the points of the bumps, with the compression of the air in the cavities giving extra buoyancy. The self-cleaning effect occurs because when dirt lands on the surface it also has few points of contact. When rain falls, the dirt adheres to the water far better than it adheres to the surface and is carried off with the water, which rolls easily over the bumps (fig. 2.4).

In Barthlott's studies, the self-cleaning effect was most noticeable in the sacred lotus (*Nelumbo nucifera*). The plant had not been easy to cultivate in Germany but when Barthlott became Director of the Bonn Botanic Garden he set about providing himself with good specimens. Around 1988, Barthlott identified the lotus as the best exponent of the art of self-cleaning; it was a magical completion of an ancient story.

Given the mythical status of the lotus it would have been reasonable to assume that the effect was peculiar to the plant, or at least to plant leaves of the lotus type. But Barthlott realized that the effect was a physical one and absolutely generic: *any* surface with bobbles of the right size, made from a water-repellent substance, would exhibit the same self-cleaning effect.

By 1988, Barthlott knew there was a technical product in view and he set out to interest the big chemical companies: 'the tribes along the Rhine', he calls them, 'those global players' (these are the major German chemical companies such as Bayer, Hoechst, BASF,

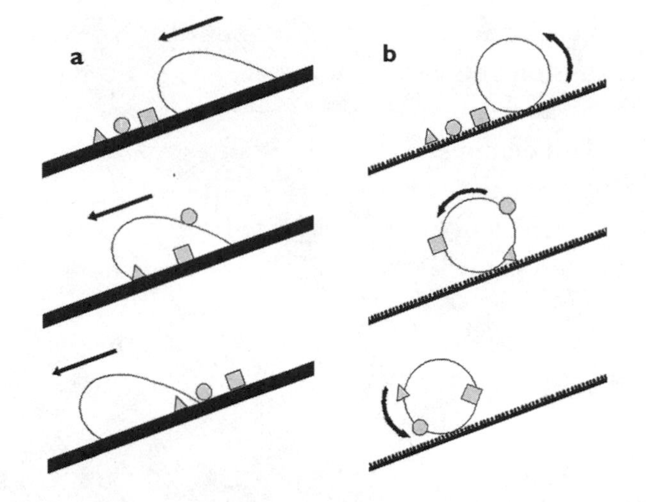

Fig. 2.4 a) On an ordinary surface dirt particles have a stronger affinity for the surface than they do for water; they remain after rainwater washes over them; b) on a Lotus-Effect® surface dirt 'sits' on top of the micro-bobbles and is easily carried off by rainwater.

Degussa). He had a party trick: he would squeeze some glue onto a leaf and show that it rolled off, leaving no trace behind. The hard-nosed industrialists refused to believe it. At first they assumed his glue was doctored and produced a tube of their own. The result was the same.

Surface-coatings specialists could not accept that they had anything to learn from plants: they said, 'Oh, it's something to do with living things.' After five years of frustration at the lack of industrial interest, Barthlott realized that he needed a technical demonstration of the self-cleaning effect, so he created the 'honey spoon', with a home-made micro-rough siliconized surface. When dipped into a honey pot, these spoons shed their entire load when tipped, leaving nothing behind (fig. 2.5). But this was a demonstration, not yet a technical product: 'It was very difficult to attach the lotus surface in a stable way, so all our home-made technical surfaces were not really intended for use. However, these first surfaces were a breakthrough: as soon as we could show them to industrial partners they were convinced. A living plant with even better properties did not have the same impact.'

Fig. 2.5 A 'honey spoon' made with a Lotus-Effect® surface. Even something as sticky as honey rolls off the spoon, leaving nothing behind.

Barthlott showed that not only could a botanist become a technical inventor but also that this botanist had fine PR antennae. He felt that the process needed something shorter and pithier to describe it than 'Self-cleaning Materials with Nanostructured Surfaces'. So, in 1992, Barthlott established the name Lotus-Effect® as a label for self-cleaning products. The lotus flower was the best example of the effect so lotus it had to be. Even so, at the time he did not realize quite how apt the name was:

> When I gave a talk to Indian students in '95 at the Humboldt Institute, they came to me afterwards and said: 'It's a symbol of purity in our religion'.
>
> I said, 'I know.'
>
> 'Do you know why?' they said. I had thought it was something esoteric – because Buddha hid under the leaves to protect himself, something like that – but no: you can find Chinese and Sanskrit poems describing the lotus, how it unfolds its leaves from dirt and muck, completely clean.

The Lotus-Effect officially entered the canon of Western inventions in July 1994 when Barthlott applied for a patent. Then, in 1997, came the classic summing up of the Lotus-Effect itself: 'Purity of the sacred lotus, or escape from contamination in biological surfaces.' This paper disclosed the Lotus-Effect in full: the biology, the physics, the implications for plant ecology and the technical possibilities. Even at this point there was resistance from some physicists to the idea of the Lotus-Effect. According to Barthlott, several journals rejected the article on the grounds that 'the so-called Lotus-Effect exists only in the imagination of the authors'. His paper concluded: 'We assume that this effect can be transferred to artificial surfaces (eg, cars, facades, foils) and thus find innumerable technical applications.'

Of course, this remark was slightly tongue-in-cheek because by now work on commercial applications was advanced; the requirements and timetables of the patent system and product development are very different to the protocols of academic publication, and anyone wishing to work in both areas simultaneously has to tread a fine line between disclosure and protecting intellectual property.

Working with Barthlott, Ispo, a paints-and-surface-coatings company, was developing a product for the exteriors of houses which, unlike existing coatings, would stay fresh and clean during its lifetime (fig. 2.6). Barthlott's patent was granted in Europe in 1998 and Ispo's paint for the exterior of buildings, Lotusan™, was launched in 1999.* It had taken 25 years from Barthlott's initial discovery to commercial exploitation. When applied, Lotusan looks like any other exterior paint. The roughness of the surface is on a scale invisible to the eye and the water-repellent silicone leaves no visible trace.

The manufacturers produce a neat demo box to demonstrate Lotusan. Half of the plates in the box are coated with Lotusan and half with a standard exterior finish of the same appearance. A bottle of distilled water and a vial of standardized fine grey ash complete the kit. The difference in properties, if not appearance, between the two surfaces is dramatic and instantly demonstrates the effect of

* Lotusan, now called StoLotusan™, is now available in the UK as a trade paint: www.sto.co.uk

highly non-wettable surfaces. Drops of distilled water on the Lotusan and non-Lotusan surfaces take on entirely different appearances. It isn't only that the former is almost spherical, with its 160° contact angle, while the other is flattened; visually, they are very different: the globule on the Lotusan surface gleams like a gem.

I opened the demo box in the company of Noah, my partner's eight-year-old grandson. When I put a drop of water on the Lotusan plate, Noah said, 'It looks like it's got sparkling water inside it' – an echo of the Japanese poet Komai's reaction in the epigraph to this chapter: 'Transforming the dew/On your life-giving leaves/Into sparkling gems!'

The other globule was dull inside because the contact angle is reversed. Multiple drops fuse instantly on the Lotusan surface; on the non-Lotusan surface, two touching globules refuse to join perfectly, a projecting pouch remaining. If you tip up the two plates, the Lotusan globule rolls off almost instantly; the other needs a slope of more than 45° to roll. The trail after the Lotusan drop is dry; a snail trail remains on the other one.

So, the water repellency is easy to demonstrate but it is the self-cleaning effect that is the commercial *raison d'être* of Lotusan. When powdered ash is scattered on both plates, a water globule cuts a swathe through the dirt on the Lotusan surface, carrying it off completely, leaving neither dirt nor water behind. On the non-Lotusan plates, the water merely smears the dirt down the plate, leaving a muddy trail.

The Lotus-Effect throws normal ideas about cleaning into disarray. You should not use detergents on Lotusan surfaces; although they do not destroy the effect, they do weaken it. The more that self-cleaning surfaces become the norm, the less cleaning agents will be used, with obvious ecological advantages. (Some of Barthlott's research documents the disastrous effect detergents can have on plant leaves, weakening their self-cleaning surfaces and laying them open to attack from moulds.)

The Lotus-Effect is the most highly developed bio-inspired technique of recent years (the all-time front-runner is the Velcro® hook-and-loop fastener – *see* Chapter 4 – but that had a 50-year head start). Lotusan is the only contemporary bio-inspired product to

Fig. 2.6 A lotus flower seen against a typical facade painted with Lotusan™ self-cleaning paint.

have made serious profits and to have achieved the distinction of being mentioned in glowing terms in company annual reports. The most difficult hurdle for bio-inspired products is not the technical development, protracted though that can be, but the crunch of coming to market and surviving the harsh reality of commercial conditions.

From its launch in 1999, Lotusan, which comes with a five-year no-cleaning guarantee, has been very successful. A measure of its success is that it is mentioned in travel guides: for example, the Nikolai-Viertel in Berlin received this write-up on www.nationmaster.com:

> The small area is famous for its traditional German restaurants and bars. Between 1997 and 1999 all houses were reconditioned (*Lotusan with Lotus-Effect*) giving this area an unmistakable touch.

Lotusan was launched at an unpropitious time for the German economy. Ispo was soon acquired by Sto, a world company with roots in Germany and America. In such a climate, even the bio-inspired paint endorsed by the purity of the sacred lotus must get its hands dirty in the commercial world. Barthlott says: 'I got the message more or less overnight that Ispo had been taken over by one of the competitors, Sto. I immediately phoned up one of our patent attorneys. He said there are two possibilities: either they want to keep it in a drawer, or they're interested in it.' They were interested in it.

The initial enthusiasm of German companies for the process has now spread beyond the country's borders – the American firm Ferro is making Lotus-Effect coatings for glass and working on coatings for metals. In Germany itself, the 'global players along the Rhine' are no longer aloof. In 2000, Barthlott took out a second patent for spray-on temporary Lotus-Effect formulations, which the chemical giant Degussa is developing.

At this point, the self-cleaning story takes an intriguing turn. There is another method of producing self-cleaning surfaces that is a mirror-image of the Lotus-Effect. Pilkington, the British company that invented the float-glass process by which most sheet glass is made, and which is licensed to every major glassmaker in the world, has developed a self-cleaning glass, Pilkington Activ™ glass, that uses a sort of anti-Lotus-Effect to achieve the same end. Instead of increasing the contact angle of water and making the surface less wettable, it *decreases* the contact angle and makes the surface *more* wettable.

The development of Activ glass is exciting and heartening for many reasons, not least for the fact that it comes not from a university department or DTI-funded start-up but from a traditional North of England manufacturing company. St Helens, Merseyside, is one of the few remaining northern towns for whom a single industry is still its calling card. You can't ignore glass and Pilkington in St Helens because, unlike so many other 'heat-and-beat' heavy industrial companies, the firm has stayed ahead of the game technically and organizationally.

The modern Pilkington stems from the 1952 invention of the float-glass process by Sir Alastair Pilkington (oddly, not a member of the founding family). The process is production-line technology *par excellence*. Glass used to be rather irregular-shaped stuff made in small quantities in unreliable furnaces. A modern float-glass factory such as the Greengate plant at St Helens can now run continuously for up to 15 years, with sand, soda ash, limestone, dolomite, sodium sulphate and recycled glass (known as cullet) feeding into a 1,600°C gas furnace at one end, a continuous ribbon of glass forming and floating on a bed of molten tin, and sheets of glass cut and stacked at the other end. The molten tin surface confers perfect flatness and the machine can be tuned to produce any desired thickness up to 20 mm.

This process produces standard raw glass, but Pilkington has now perfected a technology for depositing thin coatings on the glass from vaporized substances as it is being made; these coatings confer additional properties, as in the very common heat-insulating glass Pilkington K glass™. Activ glass is also made by this process.

When I went to see for myself, I quickly learned how important such technical advances can be to a community. In my B&B, I found that Activ glass is already famous locally and that Pilkington's share price (it had doubled in the past year) is as much a staple of conversation as the weather.

Pilkington Activ™ glass was developed at the Pilkington research centre in Lathom, 12 miles from St Helens – a green glassy haven set in parkland. Lathom is a pleasant corporate industrial environment of a kind that is increasingly rare in Britain: the calm reception area is festooned with good-employer plaques and mission statements. Simon Hurst, Pilkington Senior Technologist, wears a shirt monogrammed with both Pilkington and his own name.

Simon Hurst and Dr Kevin Sanderson, Activ's co-inventor, took me through the development process. Activ glass exploits the surprising properties of titanium dioxide, best known as the white pigment in brilliant white paints. But titanium dioxide also has unusual electro-optical properties.

The action of sunlight on titanium dioxide has the effect of charging it electrically. The charged surface then interacts with air and water vapour to create ions that can oxidize organic material. This process is called photocatalysis and it means that a titanium dioxide coating can break down any organic substance deposited on it – it is, like the lotus leaf, self-cleaning. Unlike the lotus leaf, it is strongly water-attracting, which means that water forms sheets rather than droplets on a titanium dioxide surface and if the surface is vertical or at a significant angle, water quickly rolls off, carrying away the organic material that it has degraded.

To compare the two approaches: for rain to carry off dirt particles, the dirt must have a greater affinity for the water than for the surface. This can be achieved either by making the affinity of the surface for dirt very weak – as in the Lotus-Effect – or by making the affinity of dirt for water very strong. The latter sounds less promising as water does not remove dirt easily – that is why we use soap and detergents. But the radicals produced by the action of sunlight on titanium dioxide will oxidize any organic matter (insects, pollen, plant debris, bird droppings and suchlike). Once oxidized, the organic matter dissolves in rainwater and washes away. The power of the material is constantly renewed by sunlight.

The self-cleaning ability of titanium dioxide has been known since the 1960s and in the last 10 years it has been exploited in Japan for a myriad purposes. It is used in self-cleaning tiles for bathrooms and it has medical uses – it has even been used against MRSA, the notorious multiply antibiotic-resistant *Staphylococcus* bacterium.

Ironically, titanium dioxide's photocatalytic properties were once a *problem* in its traditional use as a pigment in paints. Paints are organic materials and, of course, under exposure to sunlight the titanium dioxide attacks them. Ultraviolet light is the main cause of paint degradation in any case, but titanium dioxide was accelerating this process. The answer, as far as the paint was concerned, was to coat the titanium dioxide with silica to lock up its photocatalytic powers.

Hurst and Sanderson began to work on Activ glass in the early 1990s, and they developed a technique for coating glass when it was still very hot (about 700°C) after it has been formed on its bed of molten tin. A self-cleaning titanium dioxide layer can be applied in this way, but in any significant thickness titanium dioxide is opaque – it is, after all, a *white* pigment. The breakthrough came in perfecting this process with an ultra-thin coating, less than 20 nanometres thick. The resulting glass is perfectly transparent; next to a pane of ordinary glass it appears slightly more reflective and blue, but to all intents and purposes it is ordinary glass.

At Lathom, you feel that the world is getting better and brighter through industry. Pilkington Activ glass is the embodiment of an ancient dream: our smeary dirty world just got a little cleaner thanks to human ingenuity. And with its many cleaning properties it is a kind of miracle product.

Kevin Sanderson says, 'Activ has caught people's imagination but for many people glass is glass; we have to educate them into thinking that glass can do other things as well.' In fact, although glass may once have been taken for granted as a generic, low-profile building product, this is no longer the case. Simon Hurst says: 'Glass grows faster than GDP and has done for the last twenty or thirty years – on average four to five per cent globally every year. You've only got to look at trends in architecture – glass usage has never been higher. The new Swiss Re Tower in the City of London is entirely clad with glass.'

The final stage in the development of a technical innovation is its emergence into the real world, where it is hoped it will find a niche among 'real' people: people who have habits, customs and practices that do not respect the tidy protocols of research. Products need to be robust and easy to use to be able to claim a place 'as a dear and

Fig. 2.7 Rain forms large drops on standard window glass (foreground); on Activ™ glass it forms a continuous sheet.

genuine inmate of the household of man', as Wordsworth put it. They need to be humanized. While researching this book, a building project at my home suggested a chance to try Pilkington Activ™ glass. The small conservatory at the back of the house needed a new roof. It has a shallow 10° slope and every year it collects algae and grit that has to be laboriously cleaned off to preserve any kind of acceptable appearance. A classic potential use for Activ.

You can see the difference with Activ instantly. The conservatory roof is next to a 45° sloping glass roof at the end of the kitchen, glazed before Activ came on the market. The Activ coating gives it added reflectance that shines out against the duller standard glass. When it rains, a myriad separate drops form on the 45° standard roof but a continuous sheet quickly forms on the Activ (fig. 2.7). When there is a dew, Activ attracts it, so the water needed to do the trick is harvested from the air.

The conservatory roof sits beneath a birch tree that drops a fair amount of debris. Dry debris cannot be magically spirited away. On a roof pitched as gently as this, it needs a fairly brisk rainstorm to shift it; and fairly brisk rainstorms tend to bring down more debris. So, self-cleaning doesn't mean always clean but Activ is always at work and there is always new dirt falling. Because of the way it dries, Activ reduces spotting but doesn't entirely eliminate it. When an Activ surface does need a helping human hand, sluicing with water does the trick because nothing really sticks to this surface.

The Consumers' Association's *Which?* magazine gave Activ glass a brief write-up in June 2003. Tested against standard glass for two months, they commented:

> We struggled to find the odd smeary trace on the Activ glass. The technology doesn't work instantly, nor does it completely do away with window cleaning – you'll still need to clean the inside – and it won't deal with some marks such as paint. But it does make life simpler.

Activ glass has been on test at Pilkington since 1997 but they have simulated weathering cycles lasting much longer than that.

The prime use of Activ glass is facing out, but the omnivorous appetite of titanium dioxide for pollutants means that the technology has a potential application facing in; a case in point would be in structures suffering from 'sick-building syndrome', those large offices in which the internal atmosphere causes a sense of malaise in workers. It will help remove the oily pollutants of the kitchen, and Simon Hurst says: 'It can remove ozone – we have to be careful how we say this because of the ozone-layer, but ozone is a ground-level contaminant and Activ converts it back to oxygen.' In fact, there is a whole range of applications in the pipeline.

In many respects, the Lotus-Effect and Activ glass are equal and opposite solutions to the same problem: the road to self-cleaning can go either the super-non-wettable or super-wettable routes. The world and its materials with which we are familiar inhabit a murky zone between the two, a world in which, as the poet Philip Larkin says, 'nothing's made/As new or washed quite clean'. By discovering the extremes, we have opened up enormous possibilities: it is like extending our vision by means of infra-red and ultraviolet, radio and

X-rays – all the forms of radiation beyond the tiny band of the visible spectrum.

The success of the Lotus-Effect and Activ glass has stimulated much research and the story is far from over. The range of possible applications of the Lotus-Effect is in inverse proportion to its elemental simplicity. If the Lotus-Effect were a plant it would be seen as a rampant ecological invader. 'Superhydrophobicity' (super-non-wettability) and 'superhydrophilicity' (super-wettability) are buzz-words in many research departments.

Although the Lotus-Effect could work with any number of materials, in practice the early versions all used silicones, contemporary technology's favourite water-repellents. These are very effective but tend to be expensive. In 2003, a Turkish team of researchers found a way to make Lotus-Effect coatings from polypropylene (the stuff kitchen bowls are made of). An advantage of this simple technique is that these lotus-style polypropylene coatings can be applied to almost any material: glass, aluminium, steel, Teflon® and polypropylene itself. The only limitation is that the material the coating is applied to must not be attacked by the solvent used. Commercial exploitation of this technique is under way.

For some people, the exterior walls of their house are only slightly less remote than Alpha Centauri: self-cleaning walls are fine but if this idea is so good, can't it be used to make self-cleaning clothes? Is there any hope that in the future, accidents with red wine and coffee could be less ruinous? Yes, there is.

A self-cleaning fabric known as Nano-Care® has been developed by the American serial chemical inventor and entrepreneur David Soane (he has about 100 patents to his name and so far has started seven companies) and marketed by his firm Nanotex. Stain-resistant jeans and khakis using Nano-Care have been available in the USA from firms like Gap, Eddie Bauer and Lee Jeans since 2001 and shirts arrived a short while later. The fabrics first appeared in the UK in September 2004 with the Rocola Shirt Tec range from Morrison McConnell, a Derby-based firm and part of the Van Heusen group.

There have been many claims for stain-free clothes over the years and scepticism is understandable. The London *Evening Standard* tested them on the eve of launch by throwing lager, coffee and a particularly deep ruby red wine at the shirts. They passed: not quite every drop of the coffee was repelled but in all but the most extreme cases the shirt did what it said on the label.

The lotus leaf of Nano-Care is the peach. Peaches have a soft fuzz of hairs on the surface that function like the bobbles on a lotus leaf. They trap air and make water sit on top of the hairs. But this is very much an analogy only. If you put a peach under the tap you will see that water does run off at first, but the downy hairs are soon swamped and the surface wetted. Nano-Care whiskers are made of stronger stuff.

Nano-Care uses the lotus principle but the hairs are very tiny, less than a thousandth of the height of the lotus bumps. Compared to them, the cotton thread they stick to is an enormous tree trunk. The hairs are chemically bonded to the fibre and do not come off in the wash. And because they are so tiny, they do not change the feel of the cotton fabric appreciably.

Nanotex is a 21st-century textile company. It licenses the technology to chemical companies and buys back the nanofibre polymers to sell to textile companies which must then use the Nano-Care® trademark on the product. Nano-Care is an environmentally friendly technology in more ways than one. It makes traditional, organic cotton into a hi-tech fabric with better properties than synthetics; the process in which the nanowhiskers are attached is a normal textile process using watery solutions, and in everyday use these fabrics require fewer cleaning materials.

Whatever the technique, there will always be a need to make self-cleaning effects last longer. As Pilkington's Kevin Sanderson says:

> I think that's something that the hydrophobics have got to solve: if someone comes along and puts their fingerprint on it, it's not going to be superhydrophobic again until someone removes that smudge. The lotus leaf repairs itself because it has tiny wax crystals that grow back; if you have a surface that mimics the effect it can't do that. The Lotus-Effect is a very nice idea and it clearly works but these kinds of questions need to be answered.

The great thing about titanium dioxide is that it *is* self-renewing.

Sunlight, air and water are all it needs. Lotus-Effect paint has no such renewing power. Like all normal material surfaces, it gradually loses its powers.

Could titanium dioxide be used with Lotus-Effect coatings to produce a self-renewing capability? On the face of it this is unlikely because a waxy Lotus-Effect coating and titanium dioxide at first seem to be chalk and cheese (or oil and water). They work in opposite directions: Lotus-Effect coatings being super-water-repellent and titanium dioxide super-water-attracting. But it turns out that very small quantities of titanium dioxide can have a significant effect in breaking down organic deposits on a Lotus-Effect coating without significantly weakening its water repellency. It could so easily have been the other way round: there is an element of pure luck in technology.

Not surprisingly, nature has already combined the Lotus-Effect and Activ technology – in the shape of a beetle that lives in the Namib Desert in southern Africa. The purpose here is not self-cleaning but water collection, for this is a harsh, arid, almost rainless environment where the only moisture comes in the form of wind-driven morning fogs. Remembering that Activ glass captures the dew, gives us the clue that creatures in this environment might want to use water-attracting surfaces to harvest what water there is in fog.

This is just what the Namibian Darkling beetle does. The beetle is 2 cm long and its wing covers are warty, with bumps about half a millimetre in diameter. Under the microscope, the area between the bumps is also seen to be bumpy but at a nanoscale; the peaks of the big bumps are water-attracting whilst the rest of the surface is waxy and water-repelling.

The tips of the bumps attract and collect very fine droplets from the mist; they coalesce and grow and then the waxy portions come into play. When the droplets reach a certain size (about 5 mm), they swamp the tip and begin to roll. The other bumps help the drops roll towards the mouth of the beetle. The beetle has a rather comical 'water-collecting posture' in which it stands into the wind, face down, to present a sloping back for the water to run down.

The beetle's trick with the foggy foggy dew came to light, as so often, when researchers were looking for something else. In 2001,

Andrew Parker, a young zoologist at Oxford, came across a photograph of beetles eating a locust in the Namib Desert. The desert is probably the hottest on Earth and the locust, which had been blown there by the strong winds typical of the region, would have perished the instant it hit the sand. But the beetles were obviously comfortable.

Parker investigated the beetles, expecting to find sophisticated heat-reflection surfaces. They do indeed have such a capacity but Parker also immediately noticed the bumps on their backs. Parker is a modern researcher with an eye for bio-inspiration; the fog-harvesting ability of these beetles had been noticed back in 1976 but at the time no one looked at the mechanism. Parker immediately suspected that some adaptation of the Lotus-Effect was at work in the water-collection process.

As with the Lotus-Effect proper, you don't need a beetle, or any kind of living thing to get the effect. Water collection from fog in arid regions is an established technology: it is usually done with large nylon nets. But experiments on coated glass slides with artificial surfaces mimicking those of the beetle and control slides with entirely waxy or water-attracting surfaces quickly showed that the beetle's structure is the best for the job. Here was an efficient new way of collecting water. Parker is developing the idea with QinetiQ, the hi-tech research company spun off from the Ministry of Defence research department at Farnborough. In 2004, the process was patented and commercial applications are forthcoming.

Stripped of the needs of the beetle, the system boils down to alternating regions of water-attracting and water-repelling surfaces with the latter being the background, as it is with the beetle. The width of the water-attracting regions governs the droplet size. The technical device mimics the beetle's head-down posture by setting the collecting plates at an angle so that the water collected simply runs off into a trough. Although there is a tendency for the wind to roll the droplets back, if the size of the droplets is tuned to be large enough, they will roll against the wind into the collecting trough.

The desert-beetle water-collection mechanism is so simple and founded on such basic properties of matter it seems astonishing that it should have waited till our technology had reached such a peak of

Fig 2.8 Liquid marbles. Water drops dusted with lycopodium granules, and rendered hydrophobic with a silicone coating, form stable almost spherical 'marbles' that can even keep their shape when floating on water.

sophistication before being discovered. Water is so ubiquitous that we take it for granted. But nature exploits every possible property of a substance. Having mastered all kinds of complexity, we are now catching up on some tricks that are simplicity themselves, like this new source of water we might call Beetle Juice.

Some of the ongoing Lotus-Effect research has a playful quality in keeping with the purity of this blindingly simple idea. In 2001, two French researchers came up with a Zen-like party trick by coating drops of water so that they can roll on glass without breaking up, or even float on water itself (fig. 2.8). These 'liquid marbles' are made with lycopodium* grains coated with a silicone. This creates a lotus-like surface with almost perfect water repellency: hence their spherical shape and ability to float on water. This 'non-stick water' may eventually find applications in the packaging and delivery of

* Lycopodium is a fine yellow powder derived from the spores of the Stag's horn club moss (*Lycopodium clavatum*).

fluids, but for now it induces a Buddha-like smile at the quirkiness and eternally surprising nature of the physical world.

As usual, when we think we've invented something really far-out, nature seems to have got there first. There are aphids that, in an example of the crazily degraded lifestyles that are so common in the natural world, live all their lives inside plant galls. In fact, the galls – those warty lumps found on the undersides of tree leaves, especially on oaks – are created by the aphids, which interfere with the host plant's metabolism, thus creating the galls. In choosing, in evolutionary terms, to live like this, these aphids have created a problem for themselves. Aphids feed on the sap of plants and they produce large quantities of a whitish, sugary excrement known as honeydew. Aphids that live on the surface of plants have developed a symbiotic relationship with ants, who feed on the honeydew and protect the aphids. But gall-living aphids have no such means of disposal: they risk drowning in their own excrement unless they can easily evacuate it from the gall. The honeydew is very sticky and once an aphid gets trapped in a ball of honeydew it can't escape.

To the rescue comes super-non-wettability of an ingenious kind. The aphids produce needles that break off and line the inside of the gall with a rough waxy coating. The drops of honeydew are coated with the wax and become non-wetting honeydew parcels just like the water marbles. There is even a caste of soldier aphids whose job it is to elbow the parcelled-up honeydew balls out of the gall!

The aphid's secret was revealed in a paper, wittily entitled 'How aphids lose their marbles', by the young Indian physicist L Mahadevan and his team. Mahadevan, at Cambridge University when he did this work and now at Harvard, is one of the most dazzling figures in bio-inspiration. He is a mathematical physicist who works with biologists to unravel bio-inspired problems right across the spectrum. His papers have artistic references wherever possible, rigorous mathematics and, above all, they impart a sense of the remarkable creativity, chutzpah even, of nature in devising these solutions.

When I visited Mahadevan at Harvard, his computer desktop was a treasure trove of biological curiosities, involving origami, the draping patterns of clothes, biological springs and ratchets, and those aphids that lose their marbles. Mahadevan admits to having a short attention span, which means that he attacks these problems in a brilliant mercurial way and then passes on to the next. He is a delighted roamer in this new terrain of bio-inspiration, throwing out brilliant suggestions that others can follow up.

So we see that the Lotus-Effect is not just a matter of building maintenance. It sheds light on many strange corners of the natural world as well as adding some radiance to the built environment. Just as the self-cleaning properties of the sacred lotus were of philosophical, spiritual and artistic importance to eastern civilizations, the idea of self-cleaning can be a secular boon to the northern latitudes. In *The Poetics of Space*, the French philosopher Gaston Bachelard has suggested that cleaning might itself have spiritual/aesthetic value:

> And so, when a poet rubs a piece of furniture – even vicariously – when he puts a little fragrant wax on his table with the woollen cloth that lends warmth to everything that it touches, he creates a new object; he increases the object's human dignity; he registers the object officially as a member of the human household.

Water is one of our prime elements and in our whoring after complex chemistry we have forgotten how many subtle effects nature produces simply by manipulating water in some way. Repelling water is both the mechanism and the purpose of the Lotus-Effect, but at the nanoscale the subtle control of the water-attracting and water-repelling qualities of proteins can produce properties that have nothing to do with cleaning.

Spider silk is composed of such a protein and its strength comes from the way the fibre is spun from a watery solution, using water-attracting and water-repelling regions to create a composite structure that materials scientists would dearly love to mimic. Indeed, spider silk is regarded by many as the holy grail of materials science. The Lotus-Effect still has much scope for development but it has reached a degree of fruition: the spider guards many secrets still.

CHAPTER THREE

Nature's Nylon

> What *Skill* is in the *frame* of *Insects* shown?
> How *fine* the *Threds*, in their small *Textures* spun?
>
> RICHARD LEIGH, 'Greatness in Little'

The astounding properties of spider silk have been recognized for decades. In the force needed to break it when pulled, spider silk is about half as strong as mild steel, so the oft-quoted 'spider silk is stronger than steel' is not strictly true. Steel, however, is nearly eight times denser than spider silk so weight-for-weight spider silk *is* about six times as strong as steel. Spider silk is much more stretchy than steel, extending by 30–40% before it breaks; it is about twice as stretchy as nylon and eight times more stretchy than Kevlar®. What is special about spider silk is that it is both stretchy and tough: a rubber band will stretch more than spider silk but its breaking strength is very low. Spider silk is the only material with exceptional stretchiness *and* good breaking strength.

Spider silk has been brought to a pitch of perfection by millions of years of evolution. And this optimization means that there isn't just one generic spider silk: a single spider can make up to seven different kinds of silk, each tailored towards a specific task: the dragline from which the web is hung is the strongest, the capture threads have the greatest extensibility, and so on. Spider silk's great resilience has long suggested human applications. The web has to catch a heavy insect at speed, and bring it to a standstill without snapping and without

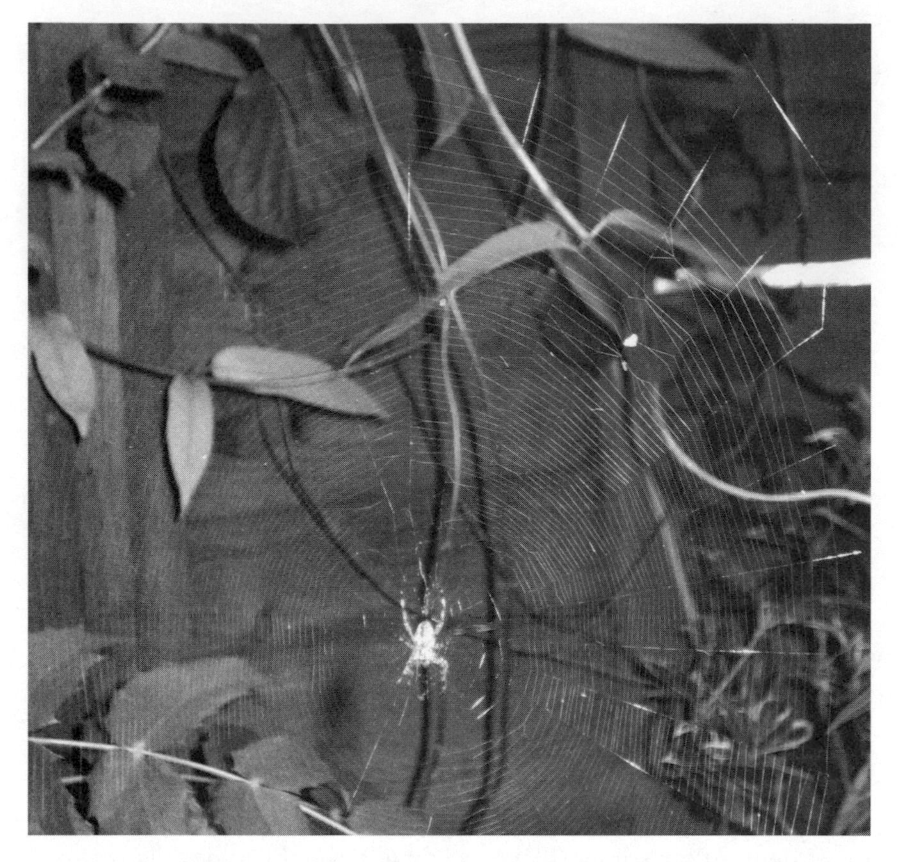

Fig. 3.1 The familiar garden spider (*Araneus diadematus*) is one of the best web spinners.

flinging it back out again in recoil, a process reminiscent of the arrester wires used to bring jets landing on aircraft carriers to a halt.

Spiders have been working their magic for over 400 million years – that's pre-dinosaur time. The oldest existing strand of spider silk was reported in 2003, preserved in Lebanese amber. It dates from the Early Cretaceous Period, more than 120 million years ago and what is fascinating about this specimen is that the small globules of 'glue' that are strung along the capture threads are still clearly visible, as they are on spider webs today.

We think of spider webs as delicate filigree structures, best seen with dew or frost accentuating their patterns. The garden spider (*Araneus diadematus*) is one of the best web spinners (fig. 3.1). But tropical spider webs can be very large: the queen of spinners is the golden orb-weaving spider (*Nephila claviceps*), which can be 5–8 cm long and 20 cm in total span: her webs are up to 2 m in diameter – big enough in fact to be useful economically. In Papua New Guinea, they have been draped across bamboo poles and looped at the end to make fishing nets. Early Western explorers also encountered such webs: in 1725, Sir Hans Sloane reported the nets were 'so strong as to give a man inveigled in them trouble for some time with their viscid, sticking quality'.

For those who fear spiders, a web large enough to enmesh a person is the stuff of nightmares. The fear of spiders is a widespread cultural phenomenon, and my interest in the silk made me question my own attitude to spiders. Primo Levi, who is always a good guide to our reactions to the natural world, summed up the symbolism of spiders like this:

> The old cobwebs in cellars and attics are heavy with symbolic significance: they are the banners of desertion, absence, decay and oblivion. They veil human works, envelop them as though in a shroud, dead as the hands which through years and centuries built them.
>
> *Other People's Trades*

There is a constellation of factors that creates a general sense of unease. There are very few large, hairy poisonous spiders (the tarantula, despite the legend, is not much more poisonous to a human being than a wasp), but all spiders, by association, share in a little of the horror that these monsters can conjure up. The spider's snaring and ambushing techniques worry some people. Something as deep-seated as spider phobia most likely has sexual connotations: the fact that the female sometimes consumes the male after mating suggests that men can associate them with women who symbolically castrate, if not devour. But women in particular are sufferers from arachnophobia. One attribute of spiders that would cause unease if it was generally known would be the fact that they have eight eyes. But most people have never seen them because they are only visible through a microscope.

Before I became seriously interested in spider silk, I had realized that not only is the web of the garden spider very beautiful but the creatures themselves, with their light speckled colouring, are much the most comely spiders you are likely to come across unless you become a dedicated arachnologist. Once I learned about the silk, my conversion to spider-worship was complete and I became ashamed of my earlier hostility.

The first documented attempt to exploit spider silk was by a Frenchman: in 1709, Xavier Saint-Hilaire Bon made gloves and stockings from the silk and presented them to Louis XIV. He wrote 'A Dissertation on the usefulness of spider silk'. The scientist René Réaumur investigated these claims in 1710 and concluded that only egg cocoon silk is good enough for spinning but it lacks lustre (a surprising finding given that all modern research focuses on the dragline silk from which the web is hung). Réaumur estimated that it would need 27,468 female garden spiders to make 1 lb of silk. Despite this discouraging report, the Chinese Emperor requested a copy of his paper and Chinese silk experts attempted to exploit spider silk. In 1876, the Chinese Emperor gave Queen Victoria a spider-silk gown. Whether this had been a century and a half in the making, since their initial interest, we do not know. Despite the impression we have of the evanescence of spider webs, the silk is durable. In Austria in the late 18th century, there was a tradition of painting on spider webs and some of these pictures exist to this day; one of them hangs in Chester Cathedral.

There is no great problem in spinning silk from a single spider. You can just about do it yourself with some improvised kit and patience. First, wait for the webs made by the garden spider, which appear in late July, then catch a spider. The spider has to be restrained, obviously, and a Styrofoam block makes a handy mount. The spider must gently be turned onto its back and a couple of very light rubber bands used to pin four legs on each side close to the body. Once the spider is stable you can start a thread by lightly touching the spinneret under the belly with a glass rod and pulling gently.

If you want to create a reel of silk, the Styrofoam block can be mounted on a piece of wood, with an improvised reel at the other end. This could be a cotton reel on a spindle with a handle. If you

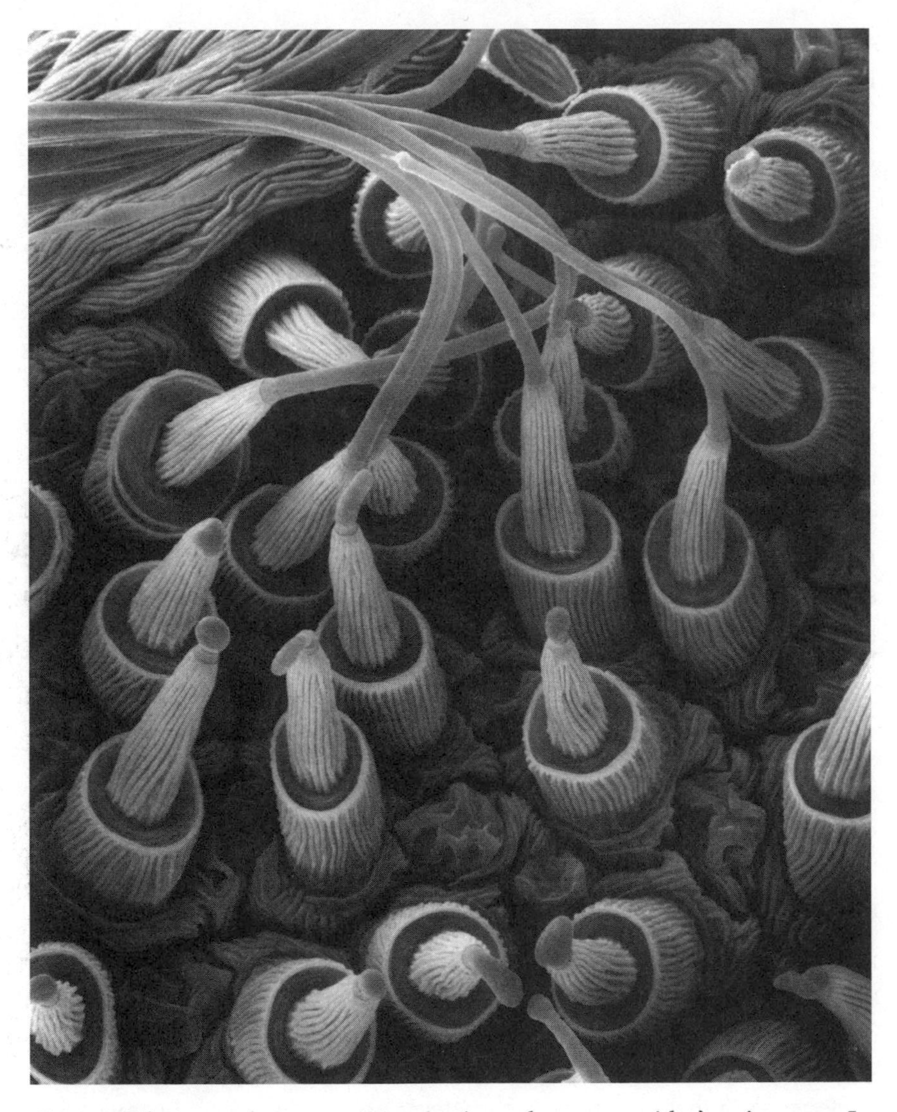

Fig. 3.2 There can be up to 600 spinning tubes on a spider's spinnerets. In action, they look remarkably like tubes of glue.

happen to know anything about hand-spinning, you could collect a few bobbins of spider silk and try to braid them into a thicker thread that would bring it closer to the dimensions of usable textiles. Whatever you do with the silk, the last stage is to let the spider go, when it will instantly return to work, repairing its web.

Although you cannot see any of this without a microscope, it is worth knowing that a garden spider has three pairs of spinnerets, each with multiple spinning tubes – more than 600 in all (fig. 3.2). Without an explanation, a picture of this apparatus might appear to be some kind of technical glue nozzle system.

The industry of garden spiders is prodigious. When I brought one in to be 'silked', its web was damaged in the process, but a few hours after releasing the spider the web had been rebuilt. Spiders keep their webs in good repair. After a few days, a web will become tatty from insect collisions, wind, dust and the spider's own movements across her domain. Every two or three days, the spider will consume the old web and build a new one, usually in exactly the same place, since they are highly territorial. Around 80–90% of a new web is protein recycled from the old one. This means that a spider catches food mainly to get the energy to build the web; it doesn't need food to supply much of the material – an example of the amazing efficiency of living processes.

Attempts have been made to silk spiders on an industrial scale. Properly set up, a single golden orb-weaver can produce 300 metres of silk in one session. The problem is that spiders cannot be farmed intensively. They are aggressive, solitary creatures who, if confined in one space, eat each other.

This naturally turns the mind towards the idea of making a synthetic silk. That this might be possible was suggested as far back as 1665 by Robert Hooke:

> Probably there might be a way found out, to make an artificial glutinous composition, much resembling, if not full as good, nay better, than that excrement, or whatever other substance it be, out of which the silkworm wire-draws his clew. If such a composition were found, it were certainly an easy matter to find very quick ways of drawing it out into small wires for use. I need not mention the use of such an invention.

For centuries the only way of making silk was with the silkworm. Archaeological evidence has shown this to be an ancient craft, going back to around 2600 BC in China. The silk moth is the only domesticated insect, having lost the power of flight, all pigmentation, and just about any desire to move or to do anything. The silk is produced by the caterpillar to cocoon the chrysalis and for this reason is not as strong as spider dragline silk. But it is a natural product that no synthetic has ever been able to match, although Japanese textile technologists have now come very close.

The basic process is as follows. The eggs are hatched and the caterpillars fed on mulberry leaves. They moult four times before they are ready to spin a cocoon in which the chrysalis will develop. The chrysalises within the cocoons are then killed by steam or fumigation. The cocoon silk consists of two filaments of the silk protein fibroin stuck together by another protein, sericin. To process the silk, the sericin is removed with hot water and the filaments drawn from water and combined to make yarn. The yarn undergoes stretching and is wound onto reels as raw silk.

Because of the finicky nature of the silkworms and the demanding cultivation regime, increasing the production of natural silk is not easy, and silk production has often been threatened by disease. In 1855, silkworms, particularly those in Europe, were afflicted by a parasitic disease called *pébrine*. This episode is the centrepiece of Alessandro Baricco's novel *Silk*, which captures in delicate prose the aura we associate with the fabric:

> He felt the lightness of a silken veil dropping onto him. And the hands of a woman – of a woman – drying him all over, caressing his skin; those hands and that material spun out of nothing. He never stirred, not even when he felt the hands move from his shoulders to his neck and the fingers –the silk and the fingers – climb to his lips and brush them once, slowly, then vanish.

Pasteur was called in to solve the *pébrine* crisis but progress was slow and this seriously focused minds on the possibility of imitating the natural process. At the time, knowledge of the chemistry of silk and all such natural substances was non-existent. Because the caterpillars grew on a diet of the leaves of the white mulberry, Count Hilaire de Chardonnet, who had worked with Pasteur on *pébrine*, tried ways of by-passing the silkworm by digesting mulberry leaves and creating a solution that could be squeezed through a nozzle similar to the silkworm's spinnerets. In fact, the main component of leaves is cellulose, a material very different to silk proteins but also a long-chain molecule. Amazingly, it did prove possible to create silk-like substances from cellulose by several processes, the best-known being rayon (1891).

The potential of silks in one of the toughest applications imaginable was realized in the late 19th century by a physician in Tombstone, Arizona: 'In the spring of 1881 I was a few feet distant from a couple of individuals who were quarrelling,' George Emery Goodfellow wrote in his diary. 'They began shooting.' Two bullets pierced the breast of one gunman, who expired from his wounds. But, on examining the body, Goodfellow found that, 'not a drop of blood had come from either of the two wounds'. He noted that 'from the wound in the breast a silk handkerchief protruded'. When he tugged on the handkerchief, it came out with a bullet wrapped inside. Evidently, the bullet had torn through the man's clothes, flesh and bones but had failed to pierce his silk handkerchief. Intrigued by this discovery, Goodfellow began to document other cases of silk garments halting projectiles – including one incident in which a silk bandanna tied around a man's neck kept a bullet from severing his carotid artery.

If silk was ever going to be used seriously for such applications it needed to be made in quantity. The mimicking of natural silks on a commercial scale began with the invention of nylon in 1937. Nylon is derived not from plant products but from very small chemical units, linked together to form long-chain molecules. Such compounds, now ubiquitous in modern civilization, are called polymers. In nylon, the link – the amide group – was the same as that in natural silks although the rest of the molecule was very different. Nylon has a much more regular structure than natural silks.

The first serious flak-jacket silk was kevlar, a tougher variant of nylon, invented in 1963. Even with nylon, kevlar and other fibres established as industrial staples, the superior properties of spider silk were alluring, but no bulk industrial or military use was pro-posed until very recently. The first serious modern application was very small scale. In the Second World War, single fibres of spider silk were used as cross-hairs for accurate range-finders – it came from black widows in the USA, garden spiders in the UK. Pioneer spider-silk researcher David Knight tells the story of the major US chemical company Du Pont, inventors of nylon and kevlar, who supplied a spider-silk sample to the US Army during the war, hoping for an order. Three years later, they politely enquired about the silk and asked whether the Army would be making an order. 'Oh, we don't need any more,' they were told, 'what you sent was fine.'

The picture changed dramatically, at least in prospect, with the arrival of genetic modification (GM) technologies in the late 1970s. In GM, a gene can be inserted into a foreign organism; the organism will function normally and produce the proteins programmed by that gene. So, in theory, if you took the gene for spider silk, and inserted it into an animal, you could make industrial quantities of silk.

Work began on this project in the 1980s and was bedevilled by nature's cussedness. Spider-silk genes turned out to be harder to handle than the insulin gene, GM's first great success. But, in June 2002, Nexia Biotechnologies in Quebec, Canada, claimed that they were able to produce industrial quantities of spider silk from the milk of genetically engineered goats. The story had a strange blend of hard military exploitation and New Age greenery. On the one hand, the US Army had been working on spider silk for many years; Nexia's silk, named BioSteel®, was developed under an Army contract for flak jackets and one of the two herds of modified goats was kept on a former United States Air Force B52 bomber base at Plattburgh, New York State. On the other hand, Nexia's President and CEO, Jeffrey Turner, waxed lyrical about this new fibre produced from meadows, goats, sun and water and spun at room temperature from a watery solution. Nylon and kevlar, the closest things we have to spider silk, are made using toxic chemicals and high temperatures and they generate toxic wastes. Turner said: 'We use water and hay; to make nylon – which has a half-life of 5,000 years [which means it's not biodegradable] – you have to sink a hole in the ground. That's not the kind of world I want to leave my kids.'

If we could manufacture large quantities of spider silk and spin it the way the spider does we would have a very special material. But 16 months on from the excited press reports of June 2002 the spider-silk story looked very different. The US Army withdrew from its collaboration with Nexia because BioSteel, as it then was, could not meet their requirements for quality or quantity.

By mid-2004, Biosteel had been downgraded even further. Development of spinning for general yarn and fabric was suspended due to the 'ongoing technical challenges of producing bulk, cost-competitive spider-silk fabrics with superior mechanical properties'. On 8 March 2005 this particular strand of the spider-silk story was fractured. Nexia's principal asset Protexia® was taken over by an american company, PharmAthene, and CEO Jeffrey Turner resigned. BioSteel® remained as the rump of a much reduced operation. So what had gone wrong? How does the spider do it and why is it so hard to emulate?

Like wool and silkworm silk, spider silk is a protein. Although it is DNA that carries the instructions for making all living things, as materials scientist Mehmet Sarikaya says: 'It is the proteins that are the workhorses in organisms. In the human body there are hundreds of thousands of different proteins; they all work at the same time in a concerted manner. You look, you eat, you think because of the interactions of these proteins.'

One of the curious things about DNA is that although it is responsible for you being a human being rather than a spider, the physical molecule is in many respects the same wherever it comes from: the arrangements of the four bases (adenine, cytosine, guanine, thymine) that make up the chain create their infinite patterns that carry the code of life, but because the bases match on comple-mentary strands – a thymine always opposite an adenine, a cytosine always opposite a guanine – the helical structure is always the same. When, in 1953, Watson and Crick deduced its structure it did not matter where the DNA came from, its X-ray crystal pattern would always have been the same. This property is essential to its function. DNA has to be as neutral and unreactive as possible: it carries the code and must not get tangled up in extraneous reactions.

Proteins, by contrast, have many varied zones of attraction and repulsion on their surfaces and they get tangled up in all kinds of ways. Protein structure is very complicated, but the long and the short is that proteins can make almost any shape you like. Many are water-soluble globular masses, like egg white, some are water insoluble and fibrous or horny, like hair or nails – or spider silk. You get some insight into protein properties every time you boil an egg. Egg white is largely made of the protein albumen. When you crack an egg, you can see that the white is clearly some kind of very thick solution. The resistance of uncooked egg white to flowing suggests that, despite its transparency, it has some filamentous structure inside it. And indeed it does. But the filaments are strongly water-loving, at least in places, and these create the familiar jelly. When you heat the egg, water is expelled from the filaments and they start to curl up. Molecules from adjacent chains meet and form cross-links. The filaments quickly tangle up into a ball and the jelly is replaced by a rubbery solid.

Although we think of food proteins and structural proteins, such as hair and wool, as radically different substances, chemically they are similar. And in fact you can turn one into the other. Milk is a foodstuff but the protein it contains, casein, can be processed to make a useful plastic. Before the advent of synthetic plastics, milk plastic was used to make small objects such as buttons.

A crude milk plastic is easy to make in the kitchen. Add a small cup of white wine vinegar to half a pint of milk in a saucepan (quantities are not crucial) and warm it, stirring constantly. Flakes of a white solid separate out. Don't boil but stir the mixture for five minutes or so over a moderate heat. Then pass through a coffee filter and wash with water to remove the vinegar. Dry the solid on kitchen roll.

The resulting product is soft and can be kneaded into a ball, or cast into shapes. As it dries, you will see how much water is trapped in the structure. The result is rather foamy, not very strong and the surface is greasy from the entrapped milk fats. To make a milk plastic for serious use, the water content needs to be controlled to allow it to escape without creating voids, the fats must be removed and the plastic hardened. Proteinaceous materials require very subtle processing to attain the right properties, but this simple experiment demonstrates that an industrial material can be made from an unlikely biological molecule. And if you can make buttons from cow's milk why not flak jackets from the spider-silk proteins in a GM goat's milk?

When still in the glands of the spider, spider silk is in jelly form and saturated with water molecules. But, like egg white, it nevertheless has a filamentous structure – in fact it is a liquid crystal. There are many kinds of liquid crystal, such as those in digital displays on watches, computers and the like, but the principle is straightforward. Although the substance in question behaves like a thick, sticky liquid, its molecules are all lined up in one direction. This affects the way the substance transmits and reflects light. In the liquid crystals used in display, a small voltage makes the molecules flip into a different orientation, thus dramatically changing their optical properties. The display consists of thousands of liquid crystal cells, appearing brighter or darker depending on the voltage applied to them, thus creating the contrast necessary for character formation. This is why the picture on a liquid crystal display disappears if you change the angle of view: the 'picture' is entirely due to the angle of reflected light.

In the case of spider silk, the liquid in the gland, known as 'dope',* has domains of strongly orientated molecules that act as liquid crystal zones. As water is removed and the acidity changes in the spinneret, the molecules become increasingly orientated, until they are aligned with the direction of flow. What emerges is a filament with the molecules strongly aligned. Pictures from the electron microscope show that the filament has a structure, rather like the large cables that hang suspension bridges: a core of bundled filaments is surrounded by a sheath made from a different protein.

Spiders spin their silks from watery solutions and it is this that so impresses commercial fibre spinners, used to working with harsh solvents. The question of proteins and water is complicated but, remembering the Lotus-Effect, it revolves around the two principles of water attraction and repulsion. Let's call them here by their technical names – hydrophilicity (water-attracting) and hydrophobicity (water-repelling). Some of the amino acids of proteins are hydrophilic, some are hydrophobic. The way these are arranged

* Dopes are solutions which when evaporated produce films or fibres. Nothing to do with drugs or silly people.

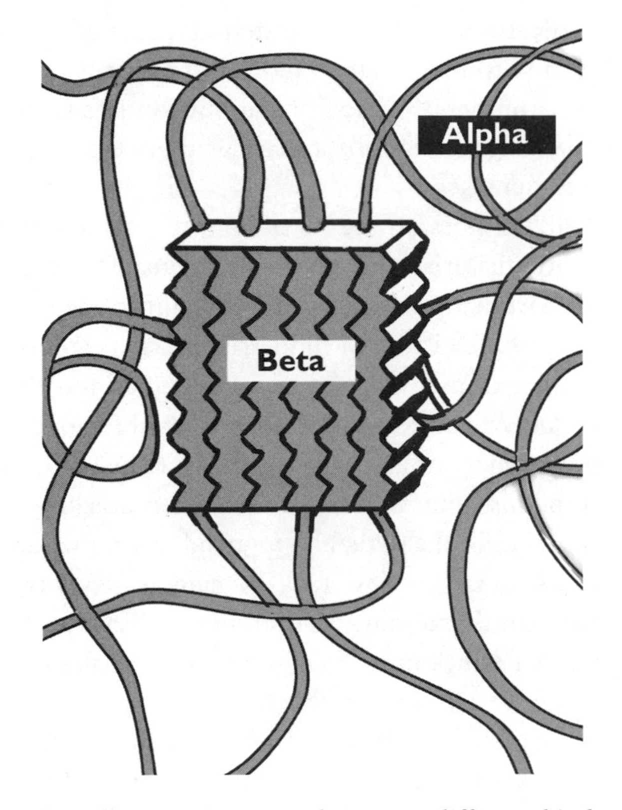

Fig 3.3 Spider silk owes its strength to two different kinds of protein domains: the hard, accordion-pleated beta sheet and the loosely coiled alpha form. These create a composite structure which is inherently tough because cracks cannot easily propagate. The principle is similar to that of fibreglass, but with the brittle and elastic elements within the same molecule.

---

along the chain affects the way the molecule folds up and how it behaves with water. Hydrophobic regions have a habit of doubling back on themselves to form a rigid crystalline pleated sheet folded like an accordion – the so-called beta sheet. The most plausible model for the structure of spider silk contains a mixture of hydrophilic portions (alpha strands) and beta-sheet crystals (fig. 3.3). This could account for the strength because it makes spider silk a composite.

A composite is a material that blends two substances with complementary properties. Fibreglass is the classic composite: glass is strong but brittle; resin is weak but stretchy. If a mat of glass has resin set around it the result is tougher and more resilient than either of the individual components. The thing about composites is that they don't crack like monolithic substances – what makes things snap is the formation of cracks.*

In a nutshell, the stress at the tip of a crack is much greater than in the rest of the material: this means that, once started, a crack is likely to travel further. In fact, it is very difficult to stop cracks once they start and a whole history of industrial disasters stems from this fact, including the Comet air crashes in the 1950s and the Hatfield rail crash in 2000. At Hatfield, minute cracks had formed on a bend where the wheel flanges of high-speed trains had been forced against the rail – something known as gauge corner cracking. Short of replacing cracked rails, the remedy is to grind down the rail until the crack disappears. So long as any cracks remain, however tiny, there is a danger. Since the Space Shuttle *Columbia*'s disaster of February 2003, the Shuttle's wings have been shown to be vulnerable to tiny cracks which can lead to catastrophic failure.

If you take a piece of fabric and give it a tug, usually nothing happens but if you make a small nick with scissors and then tug, the thing tears apart with a satisfying rip. In fact, the ripping noise signals that the failure of the material is catastrophic. This is positive feedback: the force increases as the tear lengthens and the tear lengthens as the force increases. The rails at Hatfield shattered into thousands of fragments for the same reason.

In spider silk, the brittle and the elastic components are different parts of the *same molecule* and this probably accounts for its outstanding toughness. A single molecule obviously has a greater

* There is a whole science of cracks, explained in two beautiful books by James Gordon, one of the founders of bio-inspiration: *The New Science of Strong Materials* and *Structures: or Why Things Don't Fall Down*. Gordon, who died in 1996, was a maverick engineer who had a strong belief in nature's own engineering. His books are informed by his deep love of the classical Greek world and he had an unrivalled gift for explaining technical concepts in the most down-to-earth way. Gordon's department at Reading University became the first UK biomimetics department (they call it that rather than bio-inspiration) in 1992.

structural integrity than two substances that are only physically mixed and not chemically bonded. A crystalline region in spider silk plays the role of the glass and a more elastic region plays the role of the resin.

It is one thing to recognize the superior properties of spider silk and quite another to know how to copy it. The silk exists as a fluid inside the spider and only becomes a solid filament when it emerges from the spinneret. So does the spider's secret lie in the composition of the fluid or in what happens in the spinneret?

On the face of it, it would seem most likely that it is the composition and some scientists do believe that this is the key. Clearly, this has to be significant: a material for a demanding application cannot be made out of any old stuff. But textile spinners know very well the importance of the spinning process. Nylon was not immediately recognized as a superior fibre because when it was first made in 1933 it could not be spun. It was only discovered four years later, during work on another fibre (the first polyesters), that once nylon was formed it could be toughened immensely by stretching the cold fibre. A mechanical drawing-out process within the spinneret achieves something similar for spider silk.

Thanks to genetic engineering technology, the chemical composition of spider silk is far better understood than the processes in the spinneret, and this may have accounted for Nexia's over-optimism. But even at the level of the chemistry, the spider keeps some secrets. The spider-silk protein is a very large molecule and because it does not have the precise molecular function that, say, the proteins insulin and haemoglobin do, its structure is much looser. There are many repetitive sequences and quite how precisely these have to be copied no one knows.

With a large gene such as that for spider silk, trying to make the gene work properly in a foreign organism is hard to achieve accurately: the process tends to stop short of the full chain, becoming scrambled and confused. David Kaplan, at Tufts University, near Boston, one of the researchers who has worked with spider silk from the start, explains the problems: 'Whenever you move the gene out into *E. coli* or yeast or mammalian cells, you're losing something. We spent years trying to clone the dragline silk and every time we put it into *E. coli* it would truncate down to 2.5 kD.'* Steve Arcidiacono at the US Army Soldier Center, Natick, Massachusetts, found the same thing. Steve showed me a 2.5 inch thread of GM spider silk, from *E. coli*, the first they produced – on 27 March 1998: 'It was more than we expected, but then again so was everything we accomplished given all the difficulties encountered along the way.'

Although much larger than those early synthetic spider-silk molecules, the silk molecules produced by Nexia were not full size – the complete spider-silk sequence remains hidden. In the face of the difficulties encountered by Nexia there are two main schools of thought about the way forward: one suggests that the complete protein sequence is the key – get that right and you've got it. This approach has largely been followed by the US Army team at Natick, Randy Lewis's laboratory at the University of Wyoming, and Nexia. The other approach does not deny the importance of the protein composition but stresses the chemical and physical changes that occur during the spinning process.

At Wyoming, Randy Lewis is continuing the quest for the total spider-silk structure. In October 2003 he said: 'I certainly have to believe we can completely characterize spider silks. They are just proteins in a solid form and have to conform to the features of proteins. The fact that they are in the solid state makes things more difficult but not impossible.'

David Knight, who left Oxford University's Zoology Department to found Spinox, a company dedicated to producing technical silks, has focused on the spinning process and has patented an apparatus that mimics some of the processes that occur in the spider's spinneret. David Kaplan, at Tufts University, Boston, USA, has done a lot of work on varying the conditions of silk gels before they are spun and has also produced films and sponges of reconstituted silks.

Despite all the problems, industrial spider silk is a great prize and there is a race to bring it to market. David to Nexia's Goliath is Knight's company Spinox. David Knight was an Oxford University

* *E. coli* is *Eschericia coli*, a very common bacterium that lives in the human gut and is a standard laboratory subject. 2.5 kD is the molecular weight: in native spider silk the molecular weight is as high as 300 kD.

researcher until 2003, one of the world's leading experts on spider silk, with his colleague Professor Fritz Vollrath. But, with help from Oxford University's technology transfer unit, he decided it was time to jump from the cloistered academic world into the 'nasty world out there' of venture capitalism, patents and competition; hence a tiny two-room unit on an industrial estate in Berkshire.

Not just any old industrial estate. The spider-silk story seems to be ineluctably bound up with matters military. Spinox is located at the former Greenham Common Airbase. Greenham Common was the site of a US Air Force Cruise Missile squadron in the early 1980s. It became famous for the Women's Peace Camp set up outside the base. The base was closed in 1992 following the collapse of the Soviet Union. Remarkably, the sequel is benign: the greening of Greenham Common. Although the runway still lies sinister and deserted, surrounded by barbed wire, and several of the large hangars and other buildings are still intact, the base is now a business park. Some of the units are new designer structures but many of the single-storey airbase buildings are now havens for alternative businesses.

David Knight came here in July 2003 to begin a revolution in tough fibres. Knight has a personal twist on the transformation of Greenham Common: he was last here as a member of Cruise Watch, the anti-missile protest group. He admits: 'I am a Green, or partly Green. The Green aspect [of spider silk] interests me and motivates me but we have to keep very quiet about it, because people might think that we're a bunch of Green lefties who are not really serious. We don't push it because we've realized that it's the wrong thing to say.' This highlights one of the contradictions of bio-inspiration: in the first place, the idea of spider silk has to trap a venture capitalist in its sticky threads.

So, while Nexia was burning through Canadian $2.5 million a quarter in a hi-tech gamble on genetically engineered spider silk, David Knight was pitching for a £142,000 grant from the UK Government's Department of Trade and Industry to perfect his patented spinning process.

When I met David Knight in October 2003, he was sceptical of Nexia's approach: 'Think of the cost of getting the stuff out of the milk ... they based it on the cost of ordinary milk, but this is a very valuable genetically engineered herd of goats ... They wanted us to test their material and they said: "We'd like to supply it in powder form." We'd have to use very harsh solvents to get that protein back into solution.'

David Knight's approach is different. He has worked a good deal with spiders and there are many of them in his laboratory; there are also plenty of other creatures. He sees himself in the business of spinning silk from 'amphiphilic polymers' (that is, polymers with a blend of water-attracting and -repelling properties) from whatever source.

Knight's contribution has been in his understanding of what happens inside the spider's spinneret, with water being extracted, ions exchanged, and the molecules lined up – for him, this is the key to the production of silk. He and his Oxford colleague Fritz Vollrath have patented a spinning nozzle that can handle many different solutions (fig. 3.4). It also allows for the vital ion exchange to occur while the unspun silk is still in the nozzle. In an interesting parallel with synthetic fibre spinning, the contents of a spider's spinneret are acidified as the solution passes through – and many kinds of synthetic fibres are spun from an acid bath. Fortunately, the engineering of spinning nozzles comes more naturally to David Knight than it would to some biologists: his family were carpenters and he enjoys benchtop engineering.

Knight believes that, rather than flak jackets, the first markets for spider silk are likely to be biomedical, especially fibres for closing wounds and other surgical aids: 'It's particularly appropriate because the mark-up is high, it's a market with rapid adoption of new technology, and they're eager for new product.' So how would Knight avoid the trap Nexia fell into, with BioSteel requiring time-consuming regulatory approval before it could be used in humans? He told me: 'We have discovered another silk, which is a waste product, not cocoon silk. This other silk is from a totally different source – I only realized it was a silk in the summer. This material is completely free because it is being thrown out as a by-product. I can't tell you any more but it has huge potential.' I asked Knight whether he wouldn't still have to go through the regulatory hoops, and he

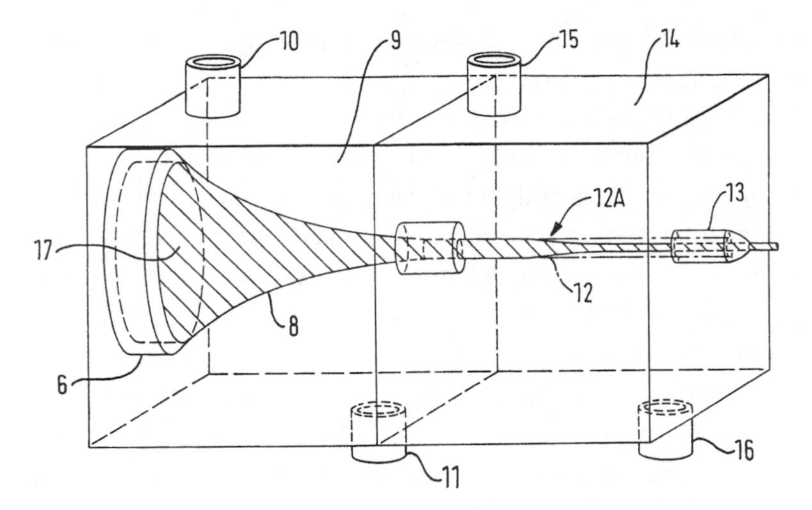

Fig 3.4 David Knight has patented a spinner that can mimic some of the processes that occur in the spider's spinneret to produce tough fibres from a range of solubilized silks.

---

replied: 'I can't tell you any more but the silk we're looking at is human.'

All the main players in spider silk became secretive at some stage in our conversations and to carry the story forward we must speculate. The main source of human fibrous protein that is thrown away in large quantities is hair. Hair is extruded slowly from the follicles, not spun rapidly like spider silk, but the structure of hair, composed of the protein keratin, is similar to silks. Silk researchers know a lot about reconstituting silks.

In 2000, Protein Polymer Technologies, Inc., a San Diego firm, took out a patent on a process for making various protein polymers, such as keratin and wool, soluble. If a good solution of keratin could be produced, Knight's patented spinner might just be the job for turning it into something special.

But hair is not the only human protein that is thrown away. While some of the blood in blood banks is used whole, many specialized blood products are derived from the rest and the inevitable waste is discarded. One blood product is the protein fibrin that turns blood into a solid scab; it is produced when fibrinogen reacts with the coagulation factor thrombin. Fibrin is already used surgically as a tissue glue and the new knowledge of protein spinning could be applied to such a protein, perhaps producing a tough fibre or film.

David Knight's final words at our interview were: 'Maybe Spinox won't succeed but somebody eventually is going to succeed with the idea of copying natural silks.' Despite Knight's affirmation, in October 2003 the spider-silk story looked none too rosy. On 31 October I learned from Steve Arcidiacono that the US Army's collaboration with Nexia had ended. Then I saw Nexia's Corporate Statement announcing their setbacks. And so, for the time being, David Knight was pursuing his mysterious human source of silk.

Nexia's problems seemed to prompt a revisionist movement amongst some spider-silk specialists. At the Materials Research Society Conference in San Francisco in April 2004, a leading forum for bio-inspired materials scientists, Christopher Viney, from the University of California, Merced, suggested that the conventional laboratory tests for strength, stiffness and toughness do not represent how spider silk behaves in the real world. One reason is that although spider silk dries and hardens when it emerges from the spinneret, it retains an affinity for water: when wet it can shrink by up to half its length and become very elastic, stretching when pulled by up to 300% against the usual 30–40%. As a result, he concluded: 'We cannot envisage natural silk serving as a long-term load-bearing material without modification. Natural silk will not rival steel as a means of suspending the deck of the Golden Gate Bridge.'

Viney has a neat way of demonstrating the effect of moisture on spider silk that you can try at home. If you surprise a spider high up it can quickly abseil to the ground on a thread. This is dragline silk. Collect a thread of at least 20 cm. Suspend it from a fixed support with a dab of super glue; the underside of the kitchen table will do. Now glue an office staple to the other end but at first support it for a couple of minutes so that the thread is not under stress. Then gently let the staple hang free. Now boil a kettle and pass the steam plume briefly across the thread a few times. Don't leave the steam billowing at the thread or it will became soaked and burdened with the weight of water.

At first the thread contracts strongly because water plasticizes the less dense regions of the chains and allows them to curl up randomly. It does not just contract, it jerks visibly. This continues with repeated steamings until the contraction is 35–40%. But then the thread starts to extend slowly, a process known as creep. With heavier weights, the initial contraction is less and the creep more rapid. Viney has discovered that microwaving spider silk in a standard kitchen microwave can greatly reduce its sensitivity to water, and this might be the way forward. Whereas, in 2002, the way ahead seemed clear, it had now become very uncertain, with new routes through the maze opening up as old ones became blocked.

Not only is water a problem but spider silk's most prized ability, that of absorbing impact, has been cast into doubt. The ability of a spider web to catch large objects such as flies without recoil suggested both the idea of the flak jacket and the aircraft carrier arrester wire. But there are obvious problems with each. Material made from fibres with 40% extension is no use for stopping a bullet: you want the back of the flak jacket not to deform at all, otherwise it is going to damage the body. The answer to this is simple: spider silk – if it could be produced in sufficient quantity – would be used in a composite with a more rigid material. As for the arrester wire, it is true that spider silk can absorb a vast amount of energy, but energy cannot be destroyed – it has to emerge somewhere. This is not a problem for the air-cooled gossamer threads of a spider web but a large, thickly coiled multi-strand arrester wire would get very hot indeed: perhaps hot enough to melt the wire.

But, in February 2004, when I met David Kaplan in his office at Tufts University, near Boston, he was remarkably bullish. As far back as 1994, Kaplan co-wrote an important paper on spider-silk protein sequences with David Knight and the two share the belief that the spider's secret is weighted in the direction of the physical drawing process by which the thread is formed.

Kaplan has worked mainly with reconstituted silks. This has been the principal method of making research-grade silk: taking silk from the one existing industrial source, the silkworm, and recycling it. Silkworm silk is not as tough as spider silk (it is intended to make cocoons so it does not have to be as strong as a web that has to withstand heavy impacts) but it can be processed in ways that bring its properties closer to that of spider silk. First, it has to be reconstituted; spun silk is redissolved in water (it needs quite harsh solvents) until it forms a solution again. It is one of nature's surprises that spider-silk 'dope' can be reconstituted in this way – it is as if you could redissolve paint once it has dried and cured.

David Kaplan's innovation is that he can now control silk spinning through the water content alone. Reconstituted silks have usually been processed using methanol to convert them from the water-soluble state into the beta-crystalline sheet. But such fibres tend to be brittle. Kaplan says: 'Now you can take the same approach but stay away from the methanol: you water-anneal them, let the water do it properly. And when you've done that you can stretch out these films 300%. They're completely stable in water and yet there's virtually no beta-sheet content ... It's really a cool material, apart from making fibres, you can make films, and stuff like the sponge you clean the kitchen sink with.'

David Kaplan was clearly excited about his work and not at all bogged down. Perhaps his techniques were the answer to the scepticism of the revisionists? He admitted that there was much he couldn't tell me because he had grants from the US Air Force and the National Institutes of Health, and he was also working with an industrial specialist in spinning silks.

The reason for David Kaplan's optimism became clear when I returned to England. The ending of the US Army's contract with Nexia did not, as I first thought, mean the end of the dream of the spider-silk flak jacket. The US Defense Department's website announces that in 2001 they awarded a Small Business Technology Transfer Program grant to David Kaplan and Foster-Miller Inc.,* a hi-tech firm in Waltham, Massachusetts, for Bio-inspired Fibers, Materials, and Properties 'in an effort to produce films from silk that possess unique and tailorable properties for emerging Air Force applications'. The grant states that: 'Ultimately, the material is likely to be well suited for highly optimized large space structures such as solar sails or space telescopes, applications where Foster-Miller is currently

* In September 2004, Foster-Miller was taken over by QinetiQ, the UK hi-tech company spun off from the Ministry of Defence research laboratories. QinetiQ is set to become a major force in bio-inspiration, with work on beetle water-collection mechanisms (*see* Chapter 2) and butterfly optics (*see* Chapter 5).

working on large deployable structures for the Air Force and NASA. In the commercial marketplace, preliminary target applications certainly include bulletproof vests as well as high-strength cords and straps, prosthetic devices, and highly abrasion-resistant textiles.'

In 2003, a follow-up grant to Foster-Miller, this time a Small Business Innovation Research Program, cited: 'Large Scale Production of Spider Silk by Immortalized Spider Cells.' Large-scale production is, of course, what Nexia had attempted with their goats. 'Immortal spiders' conjures up an image of the Arachne of legend, turned into a spider and condemned to spin for ever for presuming to be the equal of the goddess Athene. But what did it really mean in this context?

David Kaplan did not talk about immortal spiders when we met, but he did talk about stem cells. Stem cells are cells that retain the ability to develop into any kind of specialized cell, depending on which genes are switched on. Most cells can only produce daughters of themselves, but a stem cell can become any organ in the body.

Kaplan's stem-cell work is mostly aimed at developing replacement tissues, especially things like ligaments, anything made from the kind of proteins he studies: collagen and elastin. And, of course, spiders have stem cells too. Between a spider producing its tiny but exquisite filaments and goats producing spider silk in their milk, there could be a third way: to create a mass of silk-producing cells from a spider's stem cells. With Kaplan's new understanding of how to get the water out of silk dope to produce strong films and fibres, the dream of large-scale fabrication using spider silk could be realized at last. It has been a long road, but although we are not there yet, no one who has lived with the problem for the last 10 years and more seriously doubts that a spider-silk fabric will eventually be achieved.

Before I met David Kaplan, I was thinking that industrial spider silk might be 15–20 years away; but I asked him the question anyway, as you do: 'How long do you think it will take?' He replied: 'I tend to be a pretty optimistic guy and I think we're very close because we understand many of the rules we didn't understand before. One to five years I would like to say.' At the time, not knowing about the immortal spiders, I was amazed. I had heard nothing but downbeat projections from everyone else.

The spider-silk story is far from over: it has been running for about 20 years now and the materials scientists are divided between those who think that spider silk is great stuff and well worth researching in its own right and that's the end of it, and those who believe that it must inspire us to create something equally tough. It is worth remembering that for four years after its discovery nylon was thought to be useless as a technical fibre until the process of cold-drawing was discovered that dramatically increased its strength. Perhaps spider silk is still awaiting a similar boost.

In the early days of bio-inspiration, attention focused on a single outstanding attribute of a creature. The lotus has its leaves, the spider its silk, the mussel its amazing sets-under-water glue, abalone its tough composite shell – but, of course, any creature that has stayed the course during hundreds of millions of years of evolution must be versatile. Spiders are not only great web spinners, they are great climbers on any surface, especially their own webs. Like many other creatures, spiders have thousands of tiny structures on their feet that ensure good adhesion by means of forces that have only very recently begun to be understood. The fine structure of spider's feet was only investigated in 2004, after pioneering work on the feet of geckos. Each one of the spider's eight legs, it turns out, ramifies into no less than 624,000 fine hairs at the tip. The gecko has even more hairs per square millimetre than the spider, and thereby hangs an interesting tale: the bigger the creature, the more tightly packed the hairs on its feet. Why should that be?

CHAPTER FOUR

Clinging to the Ceiling

> Those rugged little bodies whose parts rise
> And fall in various inequalities,
> Hills in the risings of their surface show,
> As valleys in their hollow pits below.
>
> RICHARD LEIGH, 'Greatness in Little'

It is a lazy afternoon in the Reptile House at London Zoo: lizards lie motionless on the floor of their cell; a chameleon occasionally shifts from one twig to another. And there is a gecko – what is his idea of relaxation? He hangs on the wall on one side of his glass-box compartment, face downwards for 15 minutes or more, then moves across to do the same on the other side. It seems to cost the gecko no effort to do this; something – not his muscles – is holding him to the wall.

Geckos have always astonished everyone who has ever seen them – and that includes Aristotle, back in the 4th century BC – with their ability to run vertically up and down at will. They can scale a perfectly smooth vertical wall, even glass, and walk across a ceiling. Whether the surface is rough or smooth, wet or dry, it is all the same to the gecko.

So, what are geckos? They are a group of nocturnal lizards, about 850 species in all, found across all the southern continents and as far north as southern California, southern Europe and central Asia. The gecko on which most of the research has been done is the Tokay gecko (*Gecko gecko*), a large Asian species.

Fig. 4.1 A poster for Bob Full's talk on bio-inspired robotics.

The gecko really began to yield up its secrets in the mid-1990s in Professor Bob Full's Polypedal Lab in the Department of Integrative Biology at the University of California, Berkeley. Full is a large, ebullient stirrer and shaker at the heart of bio-inspiration today. His lab's name is apt, given the multitude of projects under his wing and the furious pedalling and promoting of the bio-inspired programme that goes on (fig. 4.1). Full is an expert on animal locomotion, which means he belongs to the biomechanic wing of bio-inspiration. Dynamics is his subject and dynamic he is, being at the centre of collaborative efforts between different disciplines and various universities. Early on he saw the potential for robotics in the way that animals move. Much of his work has focused on insect motion – the six-legged gait of creatures such as cockroaches, for instance – but the adhesion of the gecko on vertical surfaces would obviously be an attractive quality in an autonomous robot. Darpa, the Defense Advanced Research Projects Agency, which has funded so much of the work in bio-inspiration, and the firm iRobot agree. One possible use of gecko adhesion is in a climbing robot: the Mecko Gecko.

Full's approach is to find the precise principle at work in nature and then work with engineers to fabricate technical systems that do the same job. The gecko man is Kellar Autumn, who began in Full's department and is now a professor at Lewis and Clark College, Oregon.

Autumn has studied geckos for most of his professional life and his initial interest was not in their adhesion at all. Geckos are pretty remarkable all round. They are nocturnal and that means that they have to be active at low temperatures. Autumn did his Ph.D at Berkeley on the energetics of geckos – 'cold geckos running up a treadmill', as he puts it. He also spent time in Tibet, collecting geckos in their natural habitat.

The Tokay gecko is the prime gecko in every respect. It is three times more energy efficient than most creatures and its senses are very finely honed. For nocturnal hunting it has enormous eyes and it can hear the movement of an insect on a wall from across the lab. Unlike other lizards, it also vocalizes. During the Vietnam War it became notorious as the 'FU lizard'. Jumpy soldiers at night often heard a noise which sounded like the enemy taunting them in a foreign accent: 'FU, FU, FU', it screeched. The old hands never told the rookies about the creature until they had discovered it for themselves.

Autumn only started to think about the gecko's stickability when he was trying to take a break from the creature. One night, on vacation in Hawaii, he was lying on his bed when he saw a large spider on the ceiling. He was starting to think it might be dangerous when help arrived. A gecko walked across the ceiling and devoured the spider. On his return, he decided that perhaps the gecko's feet were the most interesting thing after all.

Although the gecko's ability is apparent to anyone who ever saw one, understanding it depends on being able to see the micro-structures on the pads of its feet. An unaided visual inspection offers few clues to their adhesive ability. To our eyes, the pad of the foot is crossed by transverse bands that look like a variation on standard reptile scales. Looking at the gecko's foot through an ordinary light microscope does not help much either: the pads seem to have some sort of bristly structure. It wasn't until the application of the electron microscope to the problem that the fine structure of the gecko's foot was teased out.

The electron microscope shows that there are very many bristles on the toes of a gecko: almost 500,000 on each foot. And more than that, the gecko has, as Autumn says, 'a bad case of split ends': the ends of the bristles fork into between 100 and 1,000 mini-bristles with enlarged and flattened spoon-like endings, spatulas, which, as Bob Full puts it, 'look like broccoli on the tips of the hairs'. It is these spatulas that make contact with the surface and a single gecko has about one billion of these points of contact (fig. 4.2).

So the gecko has these impressive structures, and lots of them, but how do they work? Before the era of bio-inspiration, studies of something as arcane as the gecko's foot were the province of zoologists alone. For over 100 years there have been innumerable descriptive papers dealing with gecko feet but they all came up against the limits of microscopy – the Blind Zone. Early speculations on the mechanism involved included suction and capillary force, and there were those who believed that the secret lay in the sheer number of bristle endings – if only we could see them properly.

One question that occurs to most people sooner or later is: can a dead gecko stick? Kellar Autumn tells us: 'Yes, a dead gecko can stick to vertical glass – even with a single toe. It can stay there for quite some time in fact. A large Tokay gecko died in our facility, and while it might seem morbid, we also wanted to answer this question. So we stuck it to vertical glass by a single digit, and it hung there for the

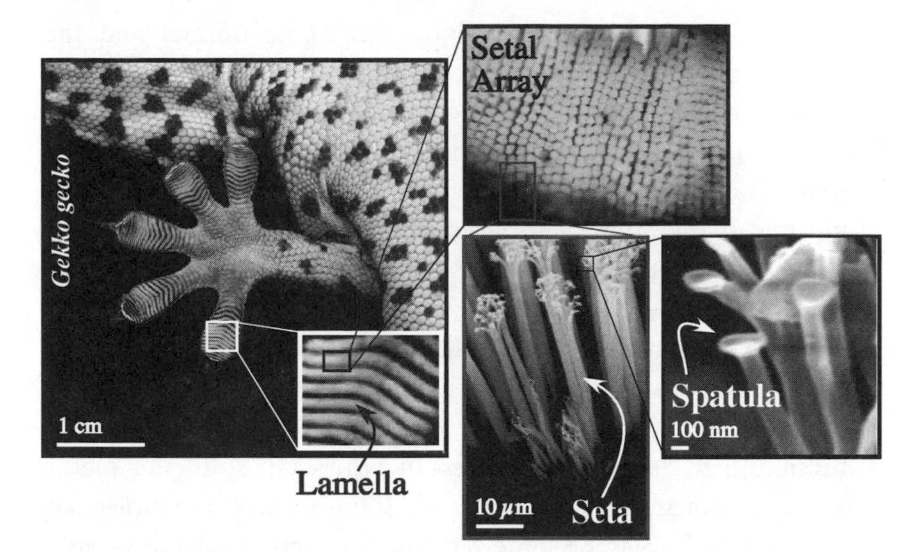

Fig. 4.2 A journey into the world of the gecko's foot, with increasing magnification at each stage until we reach the business end: the spatulas are about 200 nanometres across. A gecko has about one billion of these points of contact.

---

entire day. Indeed, if it wasn't for our concern about odour, we could have left it there much longer.'

So, the bristles on the foot are not themselves 'alive', nor do they depend on any kind of muscular activation to achieve a bond: the foot bristles are adhesive even when not attached to the gecko. Tape can be made from them, which can be harvested harmlessly from a live animal (the bristles then grow back).

In June 2000, Autumn's team published a paper in *Nature* that really brought the gecko into the age of bio-inspired solutions. They measured the adhesive force of a single gecko bristle. The experiment was very delicate and precise, given the tiny size of the bristle. This is where contemporary engineering and physics meet biology, because the micro-electronics industry has a range of techniques for working with very small structures. Bob Full's teams always include engineers, drawn from within Berkeley, Stanford and universities further afield.

Autumn's team showed that a single bristle did exhibit the force required to account for the adhesion of the foot: more than that, a single gecko bristle had *10 times* more force than you would have expected given the force exerted by the whole animal and the number of bristles. This suggested that at any one time only a fraction of the bristles was in contact. It also meant that if a gecko did have all its bristles in contact at the same time, it would be able to support a 120-kg man. In the real world it is unusual for two surfaces to touch at many points – surfaces may look to us as if they are in intimate contact but at the nanolevel hardly anything touches at all. To work effectively, the gecko has a huge margin of error. That is why it moves around so confidently and can suspend itself from a single toe: it knows it has plenty of adhesion in reserve.

But is it simply contact that explains the gecko's adhesion? Autumn likes to describe the process like this: 'The split ends merge with the surface at the molecular level.' It has to be at the molecular level, and there have to be some very small structures like those split ends to make the contact.

What is this force that acts when the gecko's bristles touch a surface, that, multiplied millions of times by the gecko's bristles, gives it its remarkable adhesion? It is a universal force of attraction that acts on all things at the bottom end of the nanorange (up to 2 nm). Further than 2 nm the force cannot be felt at all. It is not gravity, electricity or magnetism, and it is not chemical attraction. It is called the van der Waals force, after Johannes van der Waals, the Dutch physicist who first proposed these forces in 1873 to account for the temperature, pressure and volume behaviour of gases. 'Van der Waals force' sounds wonderfully obscure and highfalutin but it is just a sort of background attraction, always there but very faint. It is the Lucretian Leap again: forces on the nanoscale are different from those in the world above.

The van der Waals force only acts when two objects come within a range of under 2 nm. Most things in the real world never get close enough for this and it may not be immediately obvious why this is so. If you look at an apparently smooth surface through a scanning electron microscope that can see in detail down to 2 nm, the smoothest surface to our eyes is revealed as a ragged mountain range. Put two such 'smooth surfaces' together and in reality their area of contact will be very small. Only if a large enough area comes into contact will any adhesion be developed. The gecko's billions of spatulas can mould themselves to the contours of any surface, achieving van der Waals adhesion over a wide area. Thus it is the opposite of the Lotus-Effect: Lotus-Effect bumps are much *larger* than the gecko's bristles: they make sure that very *little* contact is made between water or dirt and the lotus surface.

Autumn's *Nature* paper concluded that the source of the gecko's fierce pull was almost certainly the van der Waals force. The only other possible candidate – not entirely ruled out – was some form of capillary attraction associated with an ultra-thin film of water on the bristles. Capillary attraction is the force that exists on the surface of a liquid: it is this that makes water curve where it meets the sides of a glass container and causes water droplets to become spherical when placed on a hard surface.

With Autumn's work, the gecko now became an interesting engineering problem, not just a biological curiosity. Professor Ron Fearing, the microfabrication engineer at Berkeley, University of California, working on dry adhesive structures modelled on the gecko, says: 'The image of what something is gets your imagination working. You *could* say: "We're going to try to make some structures that maximize the van der Waals forces." But when you see the gecko you go: "Oh, that would work, wouldn't it, a bunch of small hairs!"'

Ron Fearing is the key member of the gecko team on the engineering side, operating from Cory Hall, across the Berkeley campus from Full's lab in the Valley Life Science Building. Fearing is an engineer who became interested in biology when he realized that it is not just the nervous systems of creatures that enable them to pull off astounding physical feats: nine times out of ten it is the mechanics. As he says: 'You can get more bangs for your bucks by having a better mechanism.'

Fearing is an unusual figure in bio-inspiration, combining engineering skills at the millimetre/centimetre scale – which he uses to construct miniature flying vehicles based on the fly (*see* Chapter 7) – with the microfabrication techniques of computer-chip manufacture. From 1998, when Autumn's work was pointing strongly in the direction of a purely physical mechanism for gecko adhesion, Fearing set himself to create arrays of bristles similar to those of the gecko. To 'anyone skilled in the art', the almost alchemically

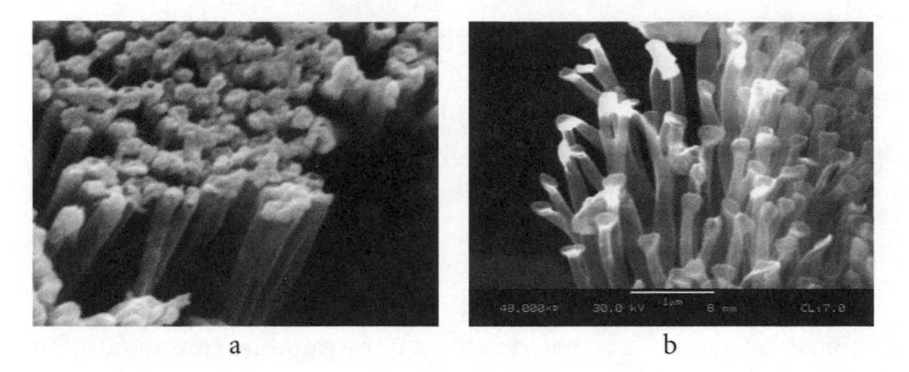

Fig. 4.3 Synthetic (a) and natural (b) gecko bristles to the same scale. The synthetic bristles have yet to match all the properties of the original but in terms of scale they are right on the button.

---

mysterious phrase that patents employ, the gecko's mechanism immediately suggests a variety of ways in which they might be fabricated.

The 2000 *Nature* paper says: 'Although manufacturing small, closely packed arrays mimicking setae [bristles] are [*sic*] beyond the limits of human technology, the natural technology of gecko foot-hairs can provide biological inspiration for future design of a remarkably effective adhesive.' This condition – 'beyond the limits of human technology' – did not last very long. It certainly did not deter Ron Fearing and he has produced some pretty good synthetic gecko arrays with the same density as the natural bristles (fig. 4.3) using ceramic moulds with nanoholes to cast polyurethane bristles.

After the paper in *Nature*, some researchers suggested that the alternative mechanism for capillary attraction needed to be tested, so Autumn's team set out to evaluate the two rival theories. This was of more than theoretical importance because if the effect depended entirely on the van der Waals force, synthetic versions could be fabricated using almost any materials and the effect would only be dependent on the size, shape and rigidity of the fibres. If capillary forces came into play, the biological processes involved were likely to be hard to reproduce. So, before investing too much in the synthetic programme it was important to be sure of the mechanism.

If water were involved in the adhesive action, different results would be expected on strongly water-repelling and water-attracting surfaces. In fact, the gecko foot bristles adhere equally well to both. These results, published in August 2002, gave the green light to attempts to fabricate artificial gecko adhesives. Only the size, shape, rigidity and, above all, the number of contact points matter.

This work caught the eye of Andre Geim, a Russian-educated physicist of German extraction, and a professor at Manchester University working in the field of condensed matter physics. He has a new purpose-built Institute of Nanotechnology and Mesophysics at his disposal, with state-of-the-art fabrication equipment in clean-rooms.* The gecko mechanism was a challenge for this kind of fabrication and he set his Russian colleagues to work on the problem. Unlike Full's team, who are committed to long-term development of their concepts, for Geim this was a project off the main line of his research interests: he simply saw an opportunity to 'prove the concept'. Geim has heretical views about the gecko's mechanism and does not believe that the case for a mechanism relying solely on van der Waals forces has been made.

The capillary forces that Autumn's team have rejected might play a role in some situations, he says: 'No one knows for sure. Of course there is capillary attraction on some surfaces. Keratin is said to be hydrophobic [water-repelling] but many people wouldn't agree with this. I don't have experience of geckos but you know from experience that your hair is hydrophilic [water-attracting] otherwise you'd never be able to take a shower. The best guess is that on this level there is no distinction between van der Waals and capillary forces. When there is water on the surface you get what I would call "water-mediated van der Waals force".'

Geim's team used an atomic force microscope to create dimples in a wax surface which was then used as a mould to create plastic pillars. In truth, these pillars were not much like the gecko's bristles. They were very short and squat by comparison. Nevertheless, even though they were far removed from the refinement of the gecko's system, the plastic pillars worked. Geim was only able to make a little of the 'gecko tape' (1 sq cm) but he extracted maximum media impact by

* Clean-rooms are dust-free facilities for the finest scale work in microfabrication.

Fig. 4.4 Andre Geim demonstrated the performance of his small sample of artificial gecko tape by getting a Spiderman toy to stick to a glass plate.

attaching it to the hand of a Spiderman toy that could easily be stuck to a horizontal glass plate (fig. 4.4).

The limitations of Geim's tape were obvious. The plastic used is rather soft and after a few uses the pillars start to stick to each other rather than to the plate. And they are water-attracting and quickly get dirty. Water-repelling, self-cleaning bristles are what is required. A gecko can reuse its bristles thousands of times on any kind of surface – rough, smooth, clean or dirty.

Although this early attempt at making a gecko adhesive was clumsy, the first transistor was also a large, lumpishly crude device, the first jet engines could barely lift a plane into the air, and the first TV pictures were barely discernible. In the history of technical inventions, proof of concept usually leads to dramatic refinements and manifold improvements in performance over a period of up to 30 years or so, when some kind of plateau is reached.

It is not enough to have the fine-scale bristles at sufficient density; enough of them need to make contact and this means that the array must be flexible in order to conform to the shape of the surface – it needs to be *compliant*, as they call it. This could be achieved by having a flexible backing, or by having flexibility in the stalks. As Ron Fearing says: 'You don't have to have soft hairs. You can make them out of hard material and make it soft by making them long and skinny. If you take a chunk of steel, and make it into a hair, it's compliant.' But if the stalks are too flexible they are going to fall over, and become entangled.

Fearing admiringly notes the ten different levels of compliance in the gecko: the spatula can bend, then the spatula hair, then the bristle hair, then the foot pad ... and so on up the leg to the gecko's body. Against the ten levels of compliance in the gecko, a standard adhesive tape has only two levels of compliance: the soft sticky surface and the flexible backing.

It is also true to say that beyond the micro-structures, the leg of the gecko has complex mechanisms that enable the sticking and unpeeling to take place. The legs of a gecko on a wall splay widely in almost a Sumo wrestler pose and the precision of its movements is very refined. This is not relevant to adhesive uses because simple peeling mechanisms suffice, but if a gecko-like robot were ever going to climb vertical walls it would need a similarly complex foot mechanism.

The gecko mechanism was patented by the Autumn/Full/Fearing team on 18 May 2004. The patent stage is always a significant and delicate one along the road to the commercialization of an invention.

Delicate, because you can neither wait till a technical process is fully realized – by then someone else will almost certainly have pipped you to the post – nor apply too early, because a patent needs to cover every possible ramification of the technique. If, for instance, a patent specifies only one method of achieving a certain end when in fact there are many, or, if very narrow conditions are cited, say, of size, reaction temperature or materials, the patent can be evaded by using a condition falling outside those specified. So a degree of legalistic comprehensiveness is necessary in a patent even though the implications of all the possibilities will not have been worked through at the time. Patents can thus contain a maze of clauses: the gecko adhesive patent application has 34 clauses in its claims section, ranging from taking bristles from living geckos to a range of fabrication techniques.

As the patent puts it: 'hundreds of thousands of setae [bristles] can be harvested without sacrificing the living being from which the setae are removed.' Autumn's team demonstrate a simple gecko sticking plaster in which three strips of gecko bristles along the tape do the job. Gecko bristles leave no gloop behind and can be used again and again, almost *ad infinitum*. The gecko, meanwhile, grows a new batch of bristles. This is, of course, strictly proof of concept: this is not the way gecko tape is going to be made in the future.

Gecko bristles make especially good sticking plasters because of the way the adhesion takes hold. The gecko pushes its foot pads against the surface and then pulls upwards. Along the line of contact, this produces the authentic gecko force (gecko tape applied to one hand would be powerful enough to stick a man to the ceiling). But if the pad is pulled away at an angle from the surface – which the gecko achieves by means of its foot-peeling mechanism – the bond is broken. In a similar fashion, when a plaster is wrapped around a finger, the pull is strong along the line of the plaster; peeling away in the traditional manner breaks the gecko bond. As Ron Fearing says: 'It pops straight off. It won't stick to the hairs, so it doesn't hurt when it comes off.'

There are already countless potential uses for gecko adhesives and no doubt there are plenty more as yet unforeseen: besides the household uses of a dry version of standard adhesive tape, a group of bristles can grip, carry and release micro-objects such as micro-electronics components; microsurgery is one more obvious application; and, because its grip is so directional, a gecko bristle can act as a clutch for micro-machines. The patent application makes a brave stab at covering the field:

> Other applications for the technique of the invention include: insect trapping tape, robot feet or treads, gloves/pads for climbing, gripping, etc., clean room processing tools, micro-optical manipulation that does not scar a surface and leaves no residue or scratches, micro-brooms, micro-vacuums, flake removal from wafers, optical location and removal of individual particles, climbing, throwing, and sticker toys, press-on fingernails, silent fasteners, a substrate to prevent adhesion on specific locations, a broom to clean disk drives, post-it notes, band aids, semiconductor transport, clothes fasteners, and the like.

The gecko story shows strong parallels with that of the Lotus-Effect. For decades everyone thought that there was something magical about the gecko, but the principle of its adhesion does not depend on the creature's behaviour, on living tissue, or on any particular chemical structure. The gecko's principle is, like the Lotus-Effect, a universal property of certain nanostructures. In fact, so universal is the gecko effect that after publication of their paper, the Berkeley team heard from many researchers in the nanofabrication field saying that they couldn't stop their nanostructures sticking to each other – and now they knew why.

In the animal kingdom, a good grip is not confined to geckos. Many creatures – beetles, flies, spiders, and other lizards besides the gecko – have excellent adhesion, due to small hairs on their feet. A team at the Max Planck Institut at Tübingen discovered a blindingly simple but apparently paradoxical rule connecting the size of the creature and the size of the bristles: the *larger* the creature the *finer the division of the bristles*. This may seem odd to us because we and other creatures that cannot walk across the ceiling have feet whose surface area grows disproportionately with increasing weight. For instance, an elephant's feet are disproportionately large compared to those of a human being, if its length alone is taken into account. This is because the weight of any creature grows with the cube of the length, whereas the area of its feet grows only with the square of the length. If a creature doubled in size, including its feet, its weight would increase by a factor of eight whilst its foot area would only have increased by a factor of four. So larger creatures have to have proportionately broader legs and feet to bear the body's weight.

But creatures that can cling to surfaces need to maximize their adhesive force, and the same rule apples to all: the more bristles, the greater the adhesion. So large creatures like the gecko, with their increased weight ration, need to have a higher density of bristles than smaller ones such as flies. A graph of the number of bristles per area against mass is that simplest of all scientific relations: the straight line. This demonstrates the great generality of nature's nano-engineering – during evolution, the density of the foot hairs of different creatures has evolved to match the weight of that creature.

The gecko's is not the first of nature's adhesive mechanisms to be useful to man. One of the earliest examples of bio-inspiration is the Velcro® brand hook-and-loop fastener. Velcro is such a good name: it is a compound of *vel*ours (the *vel-* part of the name derives from the fact that the weave is technically a velvet: a pile-woven fabric with the threads cut to produce an upstanding fringe or pile) and *cro*chet: the sound of the word mimics the combination of smooth action and raspy spikiness that is its hallmark. Almost too good a name, because the Velcro Corporation, the principal makers of hook-and-loop fasteners, are at pains to point out that the Velcro® brand is the registered name of their product and that there is no such thing as generic 'Velcro'. The company is the latest manifestation of the enterprise founded by the product's inventor, George de Mestral, and is clearly the legitimate guardian of the flame. The Velcro Corporation's naming policy is understandable but it makes discussion of their product in a book such as this difficult. But, as with the complexities of patents, protecting a trademark name is also one of the necessary complications of bringing an invention to market.

There is a persistent urban myth that the hook-and-loop fastener was a spin-off from NASA's space programme. It wasn't: NASA had no hand in its invention but astronauts did find it useful to fasten things in weightless conditions and this helped popularize the product.

The hook-and-loop fastener is a story from the pre-history of bio-inspiration and it probably evaded invention for so long because, on the face of it, it doesn't seem much of an *engineering* solution. It lacks the precision of most engineering but as a practical application that is exactly its glory. It is the first example of fuzzy logic. The zip fastener is an excellent gizmo but if the zip slips out of its notched channel it is not easy to get it back in again. But the hook-and-loop fastener doesn't have to be lined up accurately. It is what the German pioneer of bio-inspiration Werner Nachtigall* calls a 'probabilistic fastening' – a fancy way of saying that you just fumble with the thing. Whether an individual hook goes through a specific eye is irrelevant: every time you use it, enough hooks will find an eye to achieve a bond. For this reason it is considered slovenly by people who are punctilious about dress code.

As an object, the hook-and-loop fastener is delightfully anomalous. It acts and feels like a glue but it isn't sticky and is reusable thousands of times. The principle derives from the seed-propagation strategy of plants such as the cocklebur and burdock. The fruits of these plants, known as burs, are covered by thin spines with sharply hooked ends. So sharp are these that they snag anything that passes by. Animal fur is the prime target of burs but they stick to pretty well anything that comes their way.

One day in the 1940s, the target was a dog belonging to George de Mestral. Exactly what species was involved is shrouded in mystery. There are various accounts in the literature, provided by the family and the company: burdock, cocklebur and the mountain thistle have

* The German aerodynamicist Dietrich Bechert pointed out to me that many German scientists active in the field of bio-inspiration, or Bionik as it is known in Germany, have curiously appropriate names: Nachtigall (nightingale), Hummel (bumblebee), Spätz (sparrow). There's an institute for bees and the man in charge is called Bienefeld (bee-field). Then there's Professor Fisch who works on fish ... This is the phenomenon that Tom Stoppard named in his play *Jumpers* the 'Cognomen Syndrome'.

all been cited. It is not the kind of thing that gets recorded precisely in a diary ('Eureka, today I discovered Velcro') because its significance only emerges with time.

De Mestral was an inventor, trained as an electrical engineer, who lived in the family chateau near Lausanne, Switzerland. It was his pastime of hunting on the lower slopes of the Jura mountains that led to the discovery. De Mestral's grandson has said: 'It was his passion. These rare moments allowed me to be in touch with him, which was difficult. Everyone remarked that he was often lost in his own world.'

In that 'own world' of his de Mestral was on the lookout for fasteners. The story is that he had become frustrated with the difficulty of fastening the large hooks and eyes on his wife's dress before going out for the evening. De Mestral couldn't bear to be late for anything and he kept thinking that there must be a better way.

When he got back home from his walk, his dog was covered in burs and instead of just picking them off he marvelled at their tenacity, and this started him thinking. Nature maximizes the number of spines on the bur to make attachment more 'probable', in the Nachtigallian sense. The bur is spherical because it needs to maximize the angles by which it might catch a passing animal, but if the bur were rolled flat, as it were, a small square of hooks would stick to a rough fabric at whatever angle it was presented – the sort of precision docking required to fasten a single hook and loop would not be necessary.

The hook-and-loop fastener is an example of bio-inspiration from the time just before Feynman's call to consider the possibilities of nanostructures. The progression from hook-and-loop fastener to gecko adhesion shows a reduction in size from the micro-world to the nanorealm very much in line with the general thrust of engineering practice in the last 50 years. Although it is not nanotechnology, the spines on a bur are microfabrications and considerable work went into developing a viable process for manufacturing their technical equivalent.

Ideas often have to wait for a material in which to clothe them. In de Mestral's case, he was waiting for nylon – no substance before it could have been fashioned into an effective hook-and-loop fastener.

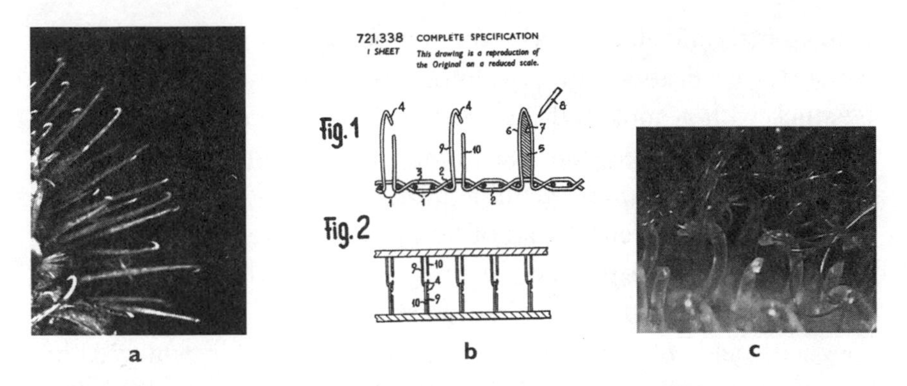

Fig. 4.5 Three stages in the gestation of the Velcro® brand hook-and-loop fastener: a) the hooked spines of a bur; b) George de Mestral's 1951 patent; c) modern Velcro under the microscope showing fibres caught on hooks.

Nylon was invented in 1937 but it was so important to the war effort that it was not available for other uses until after the war. De Mestral then took several years to find a machine process for creating hooks that could snag on the loops. To make the hooks, loops were first formed by passing nylon thread over a bar; the bar was heated to fix the shape and a knife cut the loops to produce an opening, hence a hook.

The original patent was filed in 1951 (fig. 4.5). With help from a weaver at a textile plant in Lyon, France, and a Swiss loom-maker in Basel, de Mestral perfected his hook-and-loop fastener and the product came to market in 1955.

In the original patent, the two opposing strips were more similar than they are in modern-day versions – in fact they were identical in weave but with one strip having the rows at 90° to the first, thus allowing the hooks of one row to make a secure clasp with the loops of the other. It looks as if once de Mestral tackled the engineering problem of making a reliable synthetic equivalent of the bur, for a while he forgot the lesson of the natural solution: the hooks have to have neat structures but the other half can just be tangled fur.

The American patent expired in 1978 and George de Mestral died in 1990. Hook-and-loop fasteners are now a major product worldwide and the Velcro Corporation is still the major producer. There are many variations on the original format, with some tapes using metals and able to withstand temperatures of 800°C for use in aerospace applications. In fact, almost anything that can be stuck can be stuck with a hook-and-loop fastener. It is the only bio-inspired product to have been on the market long enough to have been humanized as a 'dear and genuine inmate of the household of man'.

There is a persistent thread of humour that likes to relocate such familiar man-made products in the natural world. The television programme *Panorama* once ran an April Fools Day hoax on the spaghetti fields of Italy. The hook-and-loop fastener achieved this honorary state in a paper by Ken Umbach which reported difficulties with the Californian Velcro crop.

Three problems were encountered in the San Joaquin Valley growing area. Dry and windy conditions caused hook-and-loop spores to commingle, resulting in tangled Velcro bolls combining both strains, unprocessable by any known means. Various pests assailed the crop: the flaccidity virus weakened the hooks causing them to snap; *Millepedus miniscululus* multiplied amongst the crop until it became ensnared in the developing loops and made harvesting impossible. Finally, drought exacerbated crop-stunting salinity. Happily, by late 1996, conditions had returned almost to normal and Velcro today is blossoming. The only faint cloud on the horizon is the suspicion that the static electricity produced by billions of Velcro unzippings every day might be a factor in climate change.

Back in the real world: there is a third creature with remarkable adhesive powers, although it is better known as a culinary treat. The tenacity of mussels clinging on to rocks, ships and pier supports is legendary and they achieve this by means of an adhesive that sets under water, something human glues cannot do (it always says on the tin: 'surfaces must be clean and dry'). Unlike the gecko, mussels use a wet, sticky glue that gets into cracks in the rock and forms strong elastic cross-linkages. Not only that, but it has a chemical affinity for metals which means that it probably sticks even better to metal piers than it does to rocks. The mussel attaches itself to objects by means of a thread – the byssus – and this spreads out at the end into a plaque that sticks to the rock by means of the glue.

The mussel usually used in research into its adhesive is the blue mussel (*Mytilus edulis*), the one served in a dish of *moules marinières.* The glue has been identified as a protein but it is devilishly difficult to work with. It is so sticky from the moment it is formed that getting it to the place where you want to use it is a problem.

Nature gets round this in an ingenious way. The mussel protein is first made in an unfinished form and chemically transformed into the active glue at the last moment – one of nature's just-in-time manufacturing techniques. There are many odd things about mussel glue and one of the oddest is this last-minute transformation. The transformation creates a DOPA molecule as part of the protein chain and DOPA (dihydroxyphenylalanine) is better known as the drug used to treat Parkinson's disease and made famous by Oliver Sacks's casebook in *Awakenings* of patients who regained consciousness after decades of sleep. However, this dual use is just an interesting coincidence. DOPA has useful properties in two completely different contexts – the mammalian brain and the mussel's foot: nature is not fussed about the hierarchical divide between the two.

Inside the mussel, DOPA forms extensive cross-links, a process similar to the coagulation of egg white, but much more concentrated. The glue is remarkably water-repellent, making sure that water cannot interfere with the bond, and has a strong affinity for metal ions, hence the avidity with which mussels bind to metal piers.

Of all the processes in this book, mussel glue is closest to spider silk in being a protein nanoproduced inside living cells. Such processes, in which a cascade of chemical reactions occurs, governed by the specific structures within the cell, are the hardest natural processes to replicate. As with spider silk, genetic engineering techniques are not straightforward in this case, with cloned mussel proteins coming out shorter than those in the natural environment. And then there is the need to reproduce the mussel's trick of altering the protein after it has been synthesized to create the DOPA molecule.

There are several groups worldwide working on the processes enabling mussels to make the glue and little by little the mussel's secrets are being prised out. The key to the mussel's ability seems to be that it is a filter feeder: that is, it strains vast quantities of water to extract what nutrients it can. This means that it is able to concentrate substances such as iron that are only sparsely available in sea water. Iron is part of the mechanism of the glue, forming chemical links with DOPA. But DOPA's greed for iron is such that it latches on to it wherever it can – other metals also work to some degree. So imagine the amazing eruption into the mussels' world that man-made metal structures must have been. Mussels are attracted to iron pier-supports, bridges and ships the way moths are drawn to a flame.

In an elegant experiment, the adhesive powers of DOPA have been put to a kind of reverse use. Scientists at Northwestern University, Illinois, have used it to create a *non-stick* surface for medical applications. This is something that happens time and again in bio-inspiration. Most processes can be reversed and the reverse process can sometimes be more useful than the standard version. Some medical devices and diagnostic kits need to be protected from cells and biological fluids that would naturally stick to them. A chemical compound called polyethylene glycol (PEG) is good for this but it has to be bonded to the device. The Northwestern scientists showed that PEG can first be linked to DOPA; the DOPA then attaches to metal surfaces with its usual greed and the whole is then resistant to fouling by cells.

Meanwhile, work continues on the primary task of making a mussel glue that is easier to handle. A major use would be medical, gluing tissues together to allow healing rather than using stitches. Mussel glue is a project for the long term.

Although bio-inspired adhesives, when they reach fruition, could be used in many ways, it is more likely that the applications will be specialized and medical rather than replacements for the standard reel of sticky tape – after all, sticky tape works well and is cheap. The dryness and special peeling ability of gecko tape should prove attractive. Ron Fearing says: 'There is going to be someone out there who needs something to stick, say inside the body. They'll want something that's non-toxic and will hold for as long as they want it to and come out cleanly when they want it to.'

Although the applications listed in the gecko patent are practical, lurking at the back of the gecko story is an ancient dream: the Spiderman myth. The vertical dimension inspires a mixture of fear and desire. As with flying, some humans would like to climb buildings safely without tackle. The gecko generates such force across a small surface area that gecko gloves, provided they had a safe peeling mechanism, would bring Spiderman's feats within the realm of reality. It was Spiderman that Andre Geim thought of when he wanted to dramatize his invention, and talking to the press Kellar Autumn often says: 'Forget Spiderman, what we want is Gecko Girl.'

It is increasingly likely that gecko tape, like Lotus-Effect coatings, is going to make it. The road from proof of concept to full implementation can be a long one – often up to 20 years, and the Full/Autumn/Fearing team has been on the case since 1998. The big adhesives manufacturers are now seriously interested. If gecko development keeps pace with the example of the Lotus-Effect, we can expect commercial gecko tape around 2009, but it could be much sooner.

The gecko effect and the Lotus-Effect are both concerned with the surface of materials: the lotus presents its wax-encrusted hillocks to the world and the water rolls off; the gecko presents a rippling array of spatulas. But there are structures in many creatures that go *beneath* the surface: galleries of sculpted subterranean passages. And because they are beneath the surface they are not likely to be involved in surface effects such as adhesion or repelling water. So what are these intricate cave complexes of the nanorealm? The medium they were designed for is light and the effects nature achieves in this element – nature's iridescent lightshow – are in the vanguard of optical technology.

CHAPTER FIVE

The Gleam in Nature's Eye

> ... first-beginnings have no colour,
> But they do differ in shape, and from this cause
> Arise effects of colour variation ...
> Hues change as light fall comes direct or slanting ...
> A peacock's tail, in the full blaze of light,
> Changes in colour as he moves and turns.
>
> LUCRETIUS, *De Rerum Natura*

Fiat Lux. Light is the great glory of the world. The eye is thought to have evolved on 40 separate occasions, using 9 different mechanisms – for how could these glories not be seen? But the eye is not just passive, registering what is there: very often what is there has only evolved because some other creature can see it. Flowers evolved their gorgeous displays to attract insects; the peacock's tail evolved to attract females who, of course, had to be able to see the male in the first place.

It is believed that light became an evolutionary force over 500 million years ago, in the Cambrian era. The acceleration of evolution during this period is well known and can be explained in terms of the evolution of creatures that could perceive light for the first time, thus leading to an 'arms race' between predators and prey, the flower-and-insect pollination system, and sexual displays such as that of the peacock's tail. All these, and much more, produced the gamut of sophisticated creatures with acute senses and defences, and often stunning appearances, that we know today. Indeed, it is so like an arms race that the Pentagon has commissioned a study of the parallels between the 'arms races' of the Cambrian epoch and contemporary military systems, with a view to predicting future threats to Western security.

Light is a prime communications medium, in nature and in human technology. And although optical communications in nature might be 500 million years old, the human study and development of natural optical systems is an emerging field. Optical technology is booming, and the big prize is the promise of an all-optical computer. Computers and their silicon chip microprocessors run on electrical pulses, but could they be powered by light? And could nature's optical systems help us to do it?

The electricity in a computer is there to send pulses through logic devices that make computations and store them on disk. Light can also be pulsed like this and it is 10 times faster than electricity. And light beams can cross without interfering with each other: if electric currents cross they short circuit. (An optical computer would still need to be plugged in – the light pulses are generated by electricity in the first place.)

Light has huge advantages but it is very hard to control. Electricity threads intricate tiny pathways through silicon chips but light does not bend very willingly: it always wants to travel in a straight line. In 1987, two physicists, Eli Yablonovitch, then at Bell Laboratories, and Sajeev John at the University of Toronto, independently realized that a device could be made that would make light as easy to control as electricity. They called it the photonic crystal. It would work like the millions of transistors in a silicon chip by only allowing light of certain wavelengths to pass through: the blocked wavelengths were known as the photonic bandgap.

It is one of the joys of physics that often its greatest discoveries are not made accidentally like Fleming's chance observation of the antibiotic effect in a dish of mould left on a window sill, but as a result of purely theoretical work that predicts some previously unknown phenomenon. Radio was the classic example. James Clerk Maxwell produced the definitive theory of electricity in the 1860s and it holds to this day. From his theory he drew the prediction that

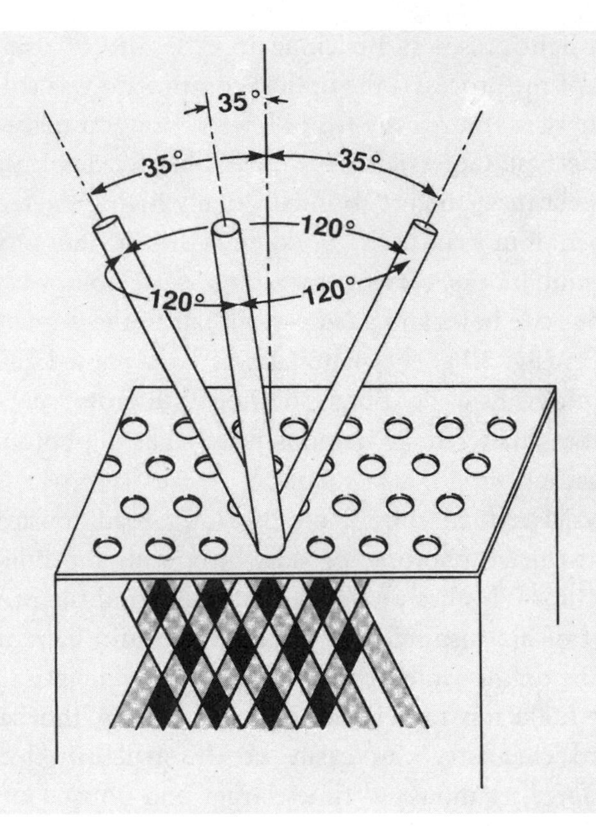

Fig. 5.1 Yablonovite, the first synthetic photonic crystal, made on a millimetre scale by drilling many holes through a ceramic.

electrical waves ought to be able to cross space and be detected by a receiving apparatus. This began the search for these waves – now known as radio – and they were first detected in 1887. Although this transmission only took place across a laboratory, by 1901 radio signals could cross the Atlantic.

Back in 1987, the photonic crystal was at the Clerk Maxwell stage. The idea of the photonic crystal was proposed at a time when optical technology was developing with the coming of the internet and there was a need for vastly increased capacity in the telephone network. The answer was fibre-optic cables that are able to carry far more information than copper wires. At present, the main trunk cables already use fibre optics but, for information travelling from one computer to another, signals have to be translated from electrical pulses into light pulses and back again. The idea of using light at every stage of the process – the optical computer – was conceived.

Many physicists across the world began trying to make photonic crystals. The first success, in 1991 by Yablonovitch himself, had a comical aspect for a subject so intellectually high-powered. He and his team spent four years at the workbench drilling millions of 6 mm holes into solid blocks, carving out a large-scale hollow crystal with Swiss cheese-style holes and a pattern similar to the crystal structure of diamonds (fig. 5.1). The material was later named Yablonovite. Nanotechnology it was not but as so often with prototypes it was the principle that mattered. It demonstrated that a photonic crystal could be made.

The physicists then started on the long road towards miniaturization, to make photonic crystals on a scale more like that of computer chips. This has proved very difficult and the problem lies at the heart of bio-inspired technology. To control light, any structure has to be on the same length scale as the wavelength concerned, just under 1,000 nm for visible light (fig 5.2). A thousand times smaller and chemistry can easily create structures by its own molecular forces; a thousand times larger and normal engineering techniques can be applied; but most photonic structures fall within the difficult nanoregion.

While Eli Yablonovitch was struggling at the workbench to make his Swiss cheese photonic crystals, many kinds of butterflies and some marine creatures, even the occasional beetle, were going about their business, as they had done for millions of years, sending optical messages by means of the very nanoscaled photonic crystals Yablonovitch and other physicists were struggling to make.

What were these creatures doing with these structures, now so prized by the physicists? Butterflies in the South American rain forest and the sea mouse, living on the almost dark ocean floor, are not very interested in high-speed optical computing; they must have their own reasons for possessing these structures.

What we notice in these creatures is the optical property we call iridescence. This is strong radiant light that changes colour as the viewing angle is changed. It is there in the peacock's tail, and in many butterflies such as the brilliant blue *Morpho* butterflies, visible a

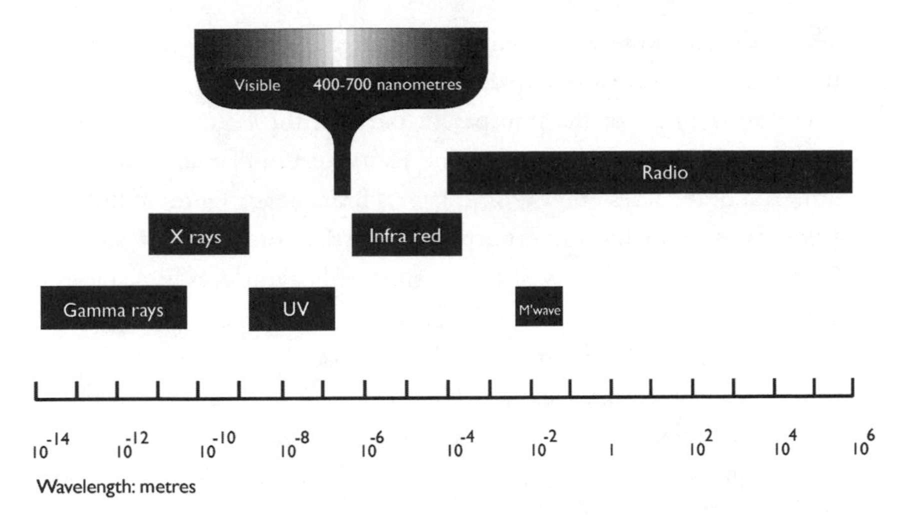

Fig. 5.2 Light is part of the electromagnetic spectrum which includes (going from small to large waves): gamma rays, X rays, ultraviolet, visible, infra red, microwave and radio. Visible light has a wavelength at the high end of the nanoscale: 400 nanometres for blue light, 700 for red. Structures on this size-scale interfere with light, causing iridescence.

---

quarter of a mile away. There is also plenty of iridescence in our everyday world: the best-known example is the compact disk, but we see iridescence dozens of times each day on credit cards, in an oily puddle in the gutter, in soap bubbles, in the gemstone opal. For most of us, the last example is not a daily experience, but it is important to our story. Opals are natural and purely mineral, but they too owe their colour to photonic crystals.

Every now and then – it happened most tellingly at the end of the 19th century – scientists start to think that some subjects are virtually complete, and then a wholly unexpected phenomenon comes into view. We find there is more to iridescence than we thought.

Lucretius's inspired guess that iridescence was caused by certain shapes (structural colour) rather than intrinsically coloured substances has long been confirmed by science. In his *Opticks* (1730) Newton wrote:

> The finely colour'd Feathers of some Birds, and particularly those of Peacocks Tails, do, in the very same part of the Feather, appear of

> several Colours in several Positions of the Eye, after the same manner that thin Plates were found to do ... and therefore their Colours arise from the thinness of the transparent parts of the Feathers; that is, from the slenderness of the very fine Hairs, or Capillamenta, which grow out of the sides of the grosser lateral Branches or Fibres of those Feathers. And to the same purpose it is, that the Webs of some Spiders, by being spun very fine, have appeared colour'd, as some have observ'd, and that the colour'd Fibres of some Silks, by varying the Position of the Eye, do' vary their Colour. Also the Colours of Silks, Cloths, and other Substances, which Water or Oil can intimately penetrate, become more faint and obscure by being immerged in those Liquors, and recover their Vigor again by being dried.

The last point Newton makes is particularly telling because it is a test applied to this day to structures suspected of producing structural colour: if the colours are altered by immersion in a fluid and regained when the fluid is expelled, it must be structural colour. Even before these mechanisms were revealed in their full glory, that the iridescent colours of butterflies and some sea creatures are produced by physical means, if not exactly common knowledge, had seeped into intellectual consciousness. Thomas Mann's *Dr Faustus* has a luminous account of this, written in 1943:

> There were many times of an evening, when Adrian's father would open his books with colourful illustrations of exotic moths and sea creatures ... The most glorious hue that they flaunted, an azure of dreamlike beauty, was, so Jonathan explained, not a real or genuine colour, but was produced by delicate grooves and other variations on the scaly surface of their wings, a device in miniature that could exclude most of the light rays and bend others so that only the most radiant blue light reaches our eyes. 'Look at that,' I can still hear Frau Leverkuhn say, 'so it's a sham?' 'Do you call the blue of the sky a sham?' her husband replied, leaning back to look up at her. 'You can't tell me what pigment produces it, either.'

When we talk about colour it is worth remembering that it is a subjective experience. Colour-blind people do not see the colours others do, although the light striking their eyes is the same. Many creatures, especially insects, can 'see' ultraviolet light, something human beings cannot do.

That there is a difference between what we perceive and what is causing the perception was first suggested by the ancient Greek philosopher Democritus some 2,500 years ago. Very little of Democritus's writings have survived but there is this tantalizing fragment:

> Ostensibly there is colour, ostensibly taste, smell, in reality only atoms and the void.

Democritus was right in believing that the phenomenon of light is not in itself coloured but not quite right about the cause (it would have been a miracle if he was: experimental science was still more than 2,000 years away and Democritus's idea was the result of reasoning alone). Colour can be said to be due to atoms and the void but only in a loose way. There are notional atoms of light – photons – but they have no weight or size: they are more like waves than particles. In physical terms, we can understand light as being waves in space, waves like the ripples on a pond. Each colour corresponds to a particular distance between successive waves.

Light is a wave motion that travels at different speeds through different media. Its speed through a vacuum is the famous 300 million metres per second and its speed through air is almost the same. But when it meets a solid but transparent medium such as glass, it is seriously slowed and this has the effect of bending the light by an angle characteristic of the medium. This is the phenomenon of refraction and the degree of bending is called the refractive index. Light is 1.5 times as fast in air as it is in glass, so the refractive index is 1.5. The refractive index is one of the keys to the optical tricks of nature and technology alike.

When light travels from a low to a high refractive index medium, the wave of reflected light is shifted by half a wavelength (a phase change). If the layer is a quarter of a wavelength thick for a light beam at a particular angle, a wave reflected from the back of the layer (reflection occurs at any surface, front or back) will have travelled

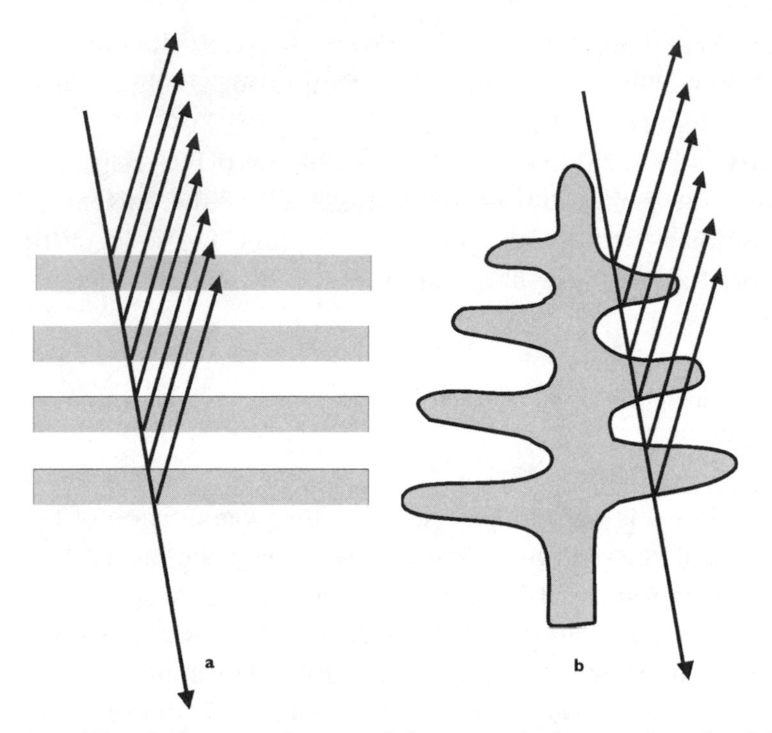

Fig. 5.3 How iridescence is created for a particular wavelength of light reflected from: a) a multilayer of differing refractive index; b) the wing scales of a *Morpho* butterfly. Light is reflected from both the front and rear surfaces and when both waves are in phase for a particular colour the reflection is greatly amplified; this causes iridescence.

half a wavelength and so be in phase with light reflected from the front. This phenomenon, known as interference, reinforces the brightness (usually for a narrow band of colours) and creates iridescence. The iridescent effect is greatly increased if a sandwich is created with alternate layers of different refractive index.

So it is distances that can be measured that lie behind our sensory experience. In the case of visible light, the rainbow colours have wavelengths of 400 nm (blue) to 700 nm (red). The colours of iridescence are caused when white light strikes anything patterned on a dimension similar to the wavelength of light. Light waves reflected from the front and back of such an object similar in dimensions to their wavelength can be perfectly in phase (fig. 5.3). This dramatically increases the brightness of that particular colour. Many creatures in nature – especially butterflies and beetles – have surface patterns with dimensions similar to the wavelength of various colours. The reason that such colours change when we shift our viewpoint or rotate the creature – often to their opposite (violet to orange; pink to turquoise) – is that the movement drastically changes the angles of rays of light from the reflecting surfaces. A ray of light at a glancing angle will have to travel much further between the elements on the patterned surface than one that strikes directly at 90°.

There is a great deal of variation between butterflies but the patterns on their wings are all variations on a standard ground plan (fig. 5.4). Ridges run along the wing scales and beneath them are hollowed-out nanocaverns. In some butterflies, structures directly beneath the ridges are patterned to reflect light, in others the cave complexes are elaborated into a foamy mass that is in effect a photonic crystal.

As with the Lotus-Effect and the gecko, the SEM was required to see the structures on a butterfly's wing that are causing the

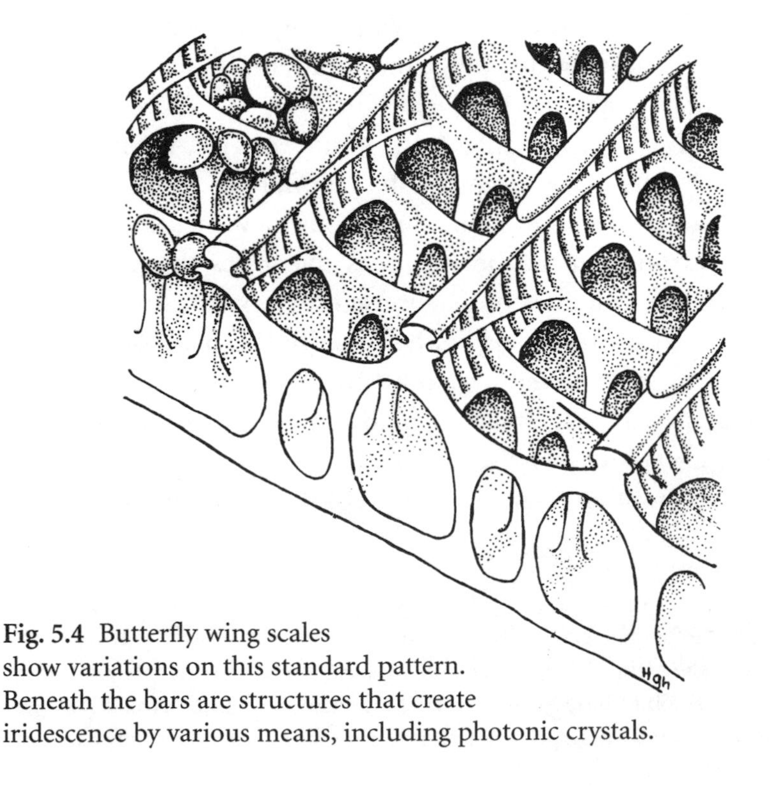

Fig. 5.4 Butterfly wing scales show variations on this standard pattern. Beneath the bars are structures that create iridescence by various means, including photonic crystals.

iridescence. From 1972, Helen Ghiradella, the doyenne of butterfly optics, at Albany University, New York, published a series of papers on butterfly wing scales with electron microscope pictures and drawings that revealed a fantastic Piranesian world of galleries and cells, all constructed to 'interfere' with light in the most dazzling ways. Here was the Lucretian underworld that produced the dazzling fireworks up above. Ghiradella said in 1991:

> The widespread distribution of iridescence throughout butterflies and moths suggests that the structures underlying the colors are easy and economical for these animals to make. Indeed, there seems to be nothing insects cannot do with their cuticle.*

What a challenge that is for materials scientists! The world can be flooded with colour just by creating miniature cavern complexes in materials of the right size range. No one with the urge to fabricate novel technical structures could fail to be inspired by these pictures. And they are rising to the challenge – but it took a while for the right people *to just look at the thing*, as Feynman urged.

Perhaps the most striking of the butterflies to show iridescence by means of interference are the bright blue *Morphos*. There are about 80 species of *Morpho*, butterflies from South America, about 150 mm in wingspan, the most dramatic being *Morpho rhetenor*. Their normal habitat is deep in the rain forest but they are often found in clearings, warming up in the sun. It is the males that are vividly coloured, so the colour is probably used for sexual display, as it is with the peacock's tail. *Morphos* have always been prized and the wings used to be collected by native peoples in Brazil to decorate ceremonial masks. Today, they are bred commercially and the wings used in jewellery and inlaid woodwork.

The structures that cause their blue colour are amongst the most

* Cuticle is the hard outer layer of all insects. It is made from chitin, which is not a protein like the fur, hair, claws, nails and hooves of mammals but a polysaccharide like cellulose, the main structural element of plants. Polysaccharides are made from sugar molecules joined head to tail in a very long chain. It is the long chain that gives them their structural properties and in this respect they *are* similar to proteins.

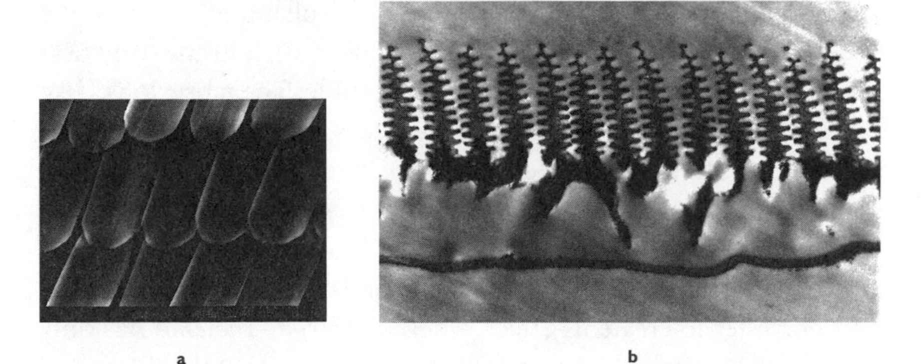

Fig. 5.5 The brilliant blue iridescence of the *Morpho rhetenor* butterfly comes from the radio-mast-like arrays on the wing scales. Seen under low magnification (a) the wing scales give no clue to these structures but there are many fine ridges running the length of the scales and beneath these ridges are the arrays (b).

---

astonishing in nature because when you see them under the electron microscope (fig. 5.5) you would swear that here is a piece of human technology. The *Morpho* wing is covered in scales that look rather like the shingles on a wooden roof. As the magnification increases you see that the scales have ridges running lengthwise down them. At higher magnifications still the full structure is revealed. The ridges have appendages that are staggered and increase in size towards the base. They look like aerials and in a sense they are. Every aspect of this intricately tooled structure has a purpose. The staggering of the side arms prevents the light being reflected only from the top layer. The alternating pattern of the arms is tuned to the wavelength of light to further increase the reflection.

The iridescence in *Morpho rhetenor* is caused by the effect of alternating ridges and air spaced at just the right distance to amplify certain colours. The effect is equivalent to that produced by continuous multiple layers of differing refractive indexes. The *Morpho*'s optical system is subtly modulated so that it always shows a strong blue from any angle, becoming only more silvery with shifting viewpoints.

Iridescence using multiple layers is the principle behind one of the first commercial applications of structural colour, named after the *Morpho* butterfly: Morphotex® fabric, introduced by the Teijin

Corporation of Japan, and developed in collaboration with the Nissan Motor Co. The fabric has already been used in the front seat covers of the Nissan Silvia Varietta Convertible. The fabric looks like a fine silk with a gentle iridescence that does not change colour with angle but gently shifts in intensity.

Morphotex is much paler than a *Morpho* butterfly and this highlights one difficulty in bringing bio-inspired products to market. Morphotex uses no less than 61 layers of alternate nylon and polyester to create its effect, whereas *Morpho rhetenor* has only 10 layers to create its richer colour. The reason is that the difference in refractive index between the two substances chosen is crucial. *Morpho* uses air and chitin, the material from which all of the external skeleton of insects is made (refractive index ratio about 1.5:1). To manufacture a fibre requires materials that are economical to produce and such a high-contrast material (relative to air) as chitin is not available to textile technologists. Nylon and polyester are similar materials so the contrast between them is much less, and many more layers are needed. If two cheap high-contrast fibre materials could be found, some dazzling textiles would result.

By the mid-1990s, while the physicists were trying to fabricate photonic crystals, biologists started to look again at other creatures under the microscope. Some of the cavern complexes in the scales of a butterfly's wings looked very like the photonic crystals Yablonovitch was trying to make. With hindsight, we can now see that after the Second World War, as instrumentation improved with the electron microscope, there were several points at which the photonic crystal and other advanced optical devices might have been discovered in nature. But, in those days, no one was looking for connections between biology and optical physics and the very *idea* of the photonic crystal was unknown. The iridescence of something that rejoices both in the humdrum common name of the sea mouse and the appropriately radiant botanical name of *Aphrodita,** had

* Although it is not perhaps as radiant as it seems. The great Linnaeus, name-giver and classifier of the whole of creation back in the 18th century, was having a joke here: in Sweden 'mouse' is slang for female genitals.

been investigated as far back as the late 1940s.

The sea mouse is a 3–4 cm long, slug-shaped creature with long iridescent porcupine-like spines. From one angle, the spines are red, from another blue-green. The colour shifts are typical of structural colour, because the hue displayed depends on the angle at which the light enters the optical structures of the creature. The iridescent fibres are hollow, with many cylindrical tubes running down their length so that the cross section looks like a slice though an Aero bar.

In the late 1990s, Andrew Parker (*see* page 51) encountered sophisticated marine optical systems while doing his Ph.D at Macquarie University, Sydney, Australia. He looked again at the iridescence of the sea mouse and with the new knowledge of photonics was able to demonstrate that the iridescent spines are photonic crystals.

The hexagonal channels running down the fibre fill with sea water and they are just so far apart that the photonic crystal effect comes into play. This is not quite the same kind of crystal that Eli Yablonovitch was trying to make. A photonic crystal can be one-, two-, or three-dimensional, and Yablonovite is a three-dimensional crystal. The sea mouse spines are one-dimensional, running down the fibre. In fact, they are very like the photonic crystal fibres invented in 1995 by Philip Russell at Bath University. His fibres were made from silica, with micro-air channels running the length of the fibre. A photonic crystal can be too perfect: if the whole fibre had channels all the way down at equal spacings, no light could be sent down the fibre. The pattern needs to be broken up to allow light in. In practice this means having a defect at the centre. At first they achieved this simply by filling in the central hole, but a better way was to have a *greater* frequency of holes at the centre. This has the dramatic effect that if white light is introduced at one end, the fibre produces green light at the other end.

These fibres were not designed for the optical computer as such but as a replacement for the standard fibre-optic cables. They are now being made commercially, although they don't yet match the standard fibre-optic cables in some respects, having losses in transmission currently 10 times greater than in conventional fibres. But Eli Yablonovitch said, in June 2004: 'I think the losses will be lower than conventional fibres within two or three years. Theoretically, it could be a lot lower.'

But if a photonic crystal fibre has to have the pattern broken to allow light to pass down it, where does this leave the sea mouse spines? Do they have such a break in the pattern? Pete Vukusic, a physicist at Exeter University who has specialized in biological iridescence, has investigated another sea worm – *Pherusa*, similar to *Aphrodita*. He says: 'It's not very different to an ordinary photonic fibre but if you go to the middle it's highly symmetric and periodic, almost bizarrely so – we kept looking in the middle for a defect and there's no defect to be found. In other words, the biological purpose is not for guidance, as we had hoped, but just for iridescence from the side.' Nature, it seemed, never found a use for wave guidance although it had developed the technology to do so. Sea mice have not yet reached the stage of being frustrated by their slow internet connections.

What the sea mouse *can* do with its iridescent spines is to signal to its own species without being seen by predators above: the colour is angle dependent and can only be seen from the side. Below 25 m, the intensity of light is only 1.5–3% that at the surface – but many creatures live below this level. Creatures that can exploit what light there is have an advantage.

The climate in optics labs in the 1990s was similar to that in the electronics labs of the 1970s. In those days, everyone could see the computer chip taking shape before their eyes and they knew that it was going to change the world. The ferment in optics – a gold rush even – brought the physicists and biologists together. Pete Vukusic says: 'Until 2000 there was very little overlap between the natural and the Yablonovitch-type photonic crystals work, synthetic photonics. It was only when we started to look at several species that we discovered that nature had beaten us to it by several million years.'

Vukusic's career is emblematic of the dramatic eruption of biology into physics. He trained as a physicist, works in the Thin-Film Photonics group in the Department of Physics, and it is clearly optical physics that he does, but his lab is full of boxes of the most gorgeous butterflies, and most of his papers, of which there have been many in the last five years, have a creature, usually – but not always – a butterfly, as their subject.

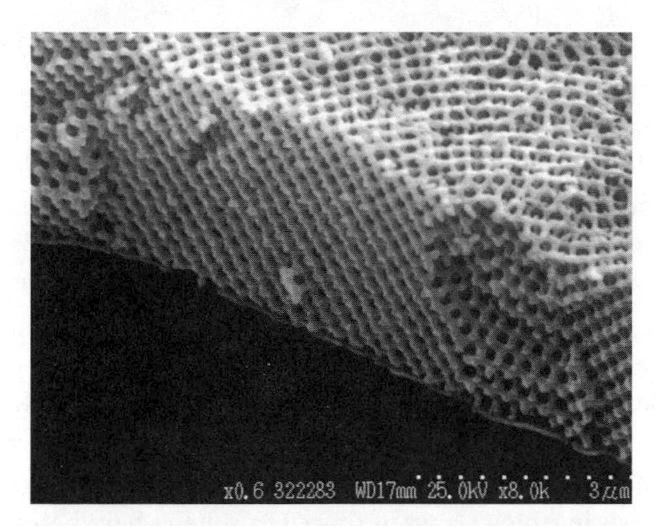

Fig. 5.6 Nature's photonic crystal. Beneath the top grid of the wing scale of the South American butterfly *Parides sosostris*, instead of the 'radio-mast arrays' of *Morpho*, there is a three-dimensional photonic crystal that produces green iridescence.

The uses butterflies make of their iridescent light are not fully understood. In general, butterflies, like all creatures, need to signal to their own kind and to avoid the attention of predators. For camouflage there are obvious advantages in a creature appearing the same colour from any angle. While most iridescence is angle dependent, which means that the butterflies are highly visible, flashing changing colours as they move, there is one structure that can produce the same colour from any angle – that is the Yablonovitch-type 3-D photonic crystal, and nature has managed to create it in several species.

In a butterfly such as the South American *Parides sosostris,* instead of the Christmas tree-style 'aerials' of *Morpho* beneath the top grid of the wing scale, there is a three-dimensional crystal: this is a photonic crystal with a diamond-type structure that allows only green iridescent light to be transmitted (fig. 5.6). The shimmering green reflection resembles the play of light on the leaves in the South American rain forest it inhabits.

Given the task of creating photonic crystals with the correct bandgap for optical computing, can these natural photonic crystals help us? Eli Yablonovitch made his first photonic crystal by the top-down approach: literally bearing down on the materials with a big drill. Ideally, we would like to be able to make nanoscale photonic crystals by methods similar to nature's, but the ease with which nature fashions the shapes described by Helen Ghiradella from materials such as chitin has yet to be emulated, although it is a live area of research.

There is one highly promising technique inspired by a natural structure that has never been alive: iridescent gemstone opal. For Shakespeare, the shifting colours of opal as they are viewed from different angles made them a symbol of inconstancy: 'thy mind is a very opal', the Clown tells the Duke of Illyria in *Twelfth Night*. Opals are minerals in which arrays of tiny silica spheres have become glued together to create an effective 3-D photonic crystal.

In geological museums and collections, besides opals, there are many fantastically shaped and coloured crystals. These have all attained their shapes through geological processes, sometimes involving heat in the depths of the earth, sometimes evaporation in pools, and sometimes great pressure as layers of minerals are forced together by moving masses of rocks. And some but not all of these processes can be duplicated in the laboratory.

In the case of the silica in opals, it naturally forms equal-sized spheres which settle into arrays in which the spheres are closely packed. This happened over geological time to produce natural opal and it can be duplicated in the laboratory. This process, of naturally falling into a regular structure, is often encountered and nature has exploited it time and time again to create her functional nano-structures. The technique is called self-assembly and it is at the heart of bio-inspiration. Instead of shaping nanopatterns by etching and cutting material away, the self-assembly approach creates conditions in which the required structures can come together using natural forces.

The structure of opal was discovered as late as 1964. Because of its simplicity it was an obvious target for synthesis and, in 1974, synthetic gem opals were first produced by the Frenchman Pierre Gilson. Gilson opals are still regarded as the best.

To make an opal, all you need is an array of silica spheres of the

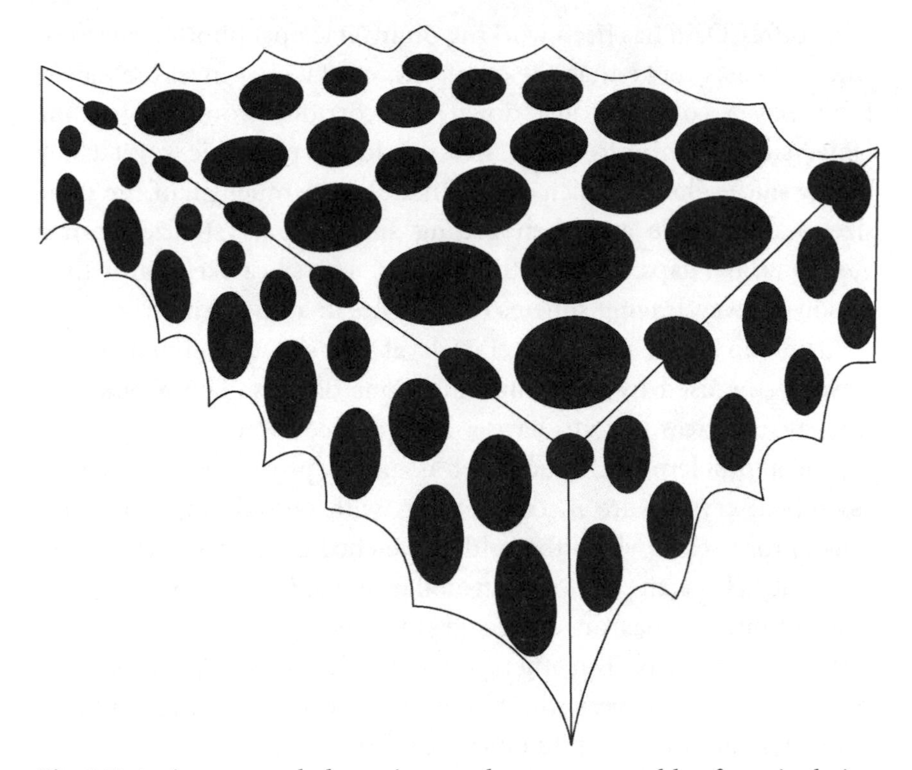

Fig. 5.7 An inverse opal photonic crystal structure capable of manipulating light in a manner similar to the way a silicon chip processes an electric current. The black holes were originally silica spheres packed together. The gaps have been filled by silicon deposited from a vapour. The silica is then chemically etched out.

right dimensions, set in a framework with a different refractive index (fig. 5.7). In natural opals, the spheres are made of crystalline silicon dioxide and the cement that holds them together is a different non-crystalline form of silicon dioxide. Although chemically the same, these two physically different forms have sufficiently contrasting refractive indexes to produce the opal effect.

In opals, both natural and synthetic, it is the contrast in refractive index between the silica and the air voids within them that creates the iridescence: reverse the spheres and voids – this is called an inverse opal – and it is still a photonic crystal. This is the sort of thing that gets materials scientists excited because templating reactions easily produce inverse structures.

Geoffrey Ozin has been working on inverse opal photonic crystals since he discovered that the photonic crystal's co-inventor Sajeev John was a colleague just down the corridor from him at the University of Toronto. Ozin has produced photonic crystals by letting silica spheres, which can be made pretty much all of the same size, self-assemble and then getting silicon to crystallize from a vapour in the gaps between the spheres. The silica skeleton is then dissolved away, leaving spheres of air holes in a silicon matrix.

Ozin can make photonic crystals at the size demanded by the wavelengths used by telecommunications devices. The problem is that such devices have to have a high degree of accuracy and this is still a problem with such wet-assembly processes. Ozin's self-assembled crystals are in competition with crystals made by top-down processes in which the voids are etched by laser. But the prizes are great. 'This is the next trillion-dollar industry,' says Gerald Lynch, who admittedly has an axe to grind – he is CEO of Photonics Research, Ontario. But there are sceptics and, 15 years after Yablonovitch's discovery, the photonic equivalent of the computer chip – reliable and cheap to fabricate – remains elusive.

Although there is still a lot to learn about how butterflies, sea mice and other marine creatures use their powers, being visible to the right individuals and invisible to the wrong is clearly a large part of the struggle for life. One of the deepest things we share with the natural world is the need to display and conceal. At our most instinctive level we are creatures who need to be heard above the mass, or sometimes to shrink into the shadows, to woo and to wow, to make a killing. Appearances, in the struggle for life, are more than superficial.

Hence, one of the most important applications of bio-inspired optical technology may be in anti-counterfeiting devices. The codes we use to establish our bona fides need to be resistant to copying. Patterns that use some of the light-bending properties exploited by the butterflies are much harder to copy than conventional visual patterns.

We already use optical devices – holograms on credit cards, even iridescent banknotes in some countries – but these are in early stages of sophistication. Identity theft is one of the fastest growing crimes and forgery is becoming the criminals' favourite activity. Paper documentation is less secure than ever, with fake degree certificates easily available on the internet. In the long run, any technology is copiable but some advanced technologies can put our key validation processes beyond the practical reach of counterfeiters. Perhaps there is a lesson to learn from the way natural creatures signal only to those who should receive the messages?

One of the butterflies' best strategies is to use polarized light. Polarized light is light that vibrates in one plane only. If you imagine a light wave as a kink passing down a lasso, in normal light the kink wobbles 360° all around the direction of travel; in polarized light it travels in one plane only. If a butterfly's wings produce polarized light at a certain angle and its own species is sensitive to that angle of polarized light, the chances are that other species won't be. So polarized light can act like a code between members of a secret society. As yet, little work has been done on this but recently polarized light was shown to be used for finding mates in *Heliconius* butterflies.

Some of the butterflies' iridescent mechanisms may lead to new products. *Ancyluris meliboeus* creates a flickering effect with a different kind of wing scale structure to that of *Morpho*. *Ancyluris* has coloured patches under the wings that show a broad range of colour from orange to blue, but for a 60° arc there is no colour at all – there is a dark zone. It achieves this by having its ridge scales angled at 30° to the base of the scales. The extreme dependence of this iridescence on angle makes it an attractive technical proposition and the Exeter team, together with QinetiQ, the hi-tech research company, has taken out a patent on the system. It could be used for document security, decorative packaging, advertising logos or textile fibres.

The butterflies' iridescence is a sophisticated form of reflection and simpler forms of reflection and anti-reflection are also used in both nature and engineering. The most famous bio-inspired reflector is the Catseye®. Whether in fact it *was* bio-inspired is a moot point: there are several accounts of its genesis. The real cat's eye has a special layer, the tapetum, which reflects back light unabsorbed by the retina to increase the illumination in low light. The characteristic glow this gives a nocturnal cat may or may not have influenced Yorkshireman Percy Shaw in his invention. The connection between the invention and the original was certainly noticed soon after – hence the evocative trade name.

Shaw ran a car repair business, and this brought home to him how dangerous unlit country roads were at night. In a television interview in 1968, he claimed that he saw a reflecting road sign on a foggy night and thought: 'We want those things down on the road, not up there.' He patented the first road reflectors in 1934 but they didn't really come into their own until the blackouts of the Second World War, when headlights had to be masked. It was realized that the reflectors threw no light upwards where enemy planes might spot them and so were blackout-approved. This, of course, is the same principle employed by the sea mouse to signal to its own species without betraying itself to predators above, although no one was much concerned with analogies like this at the time.

Shaw founded Reflecting Roadstuds Ltd in his home town of Halifax in 1935 and the company makes Catseyes to this day. Shaw was a classic self-made eccentric English inventor. He became rich but lived the whole of his bachelor life in the same house, allowing himself the luxury of two Rolls-Royces, from one of which he would alight at the local fish and chip shop to eat his meal in the back of the car.

Catseyes are beautifully simple devices (fig. 5.8). They reflect light only where it is needed and they are also self-cleaning – 65 years before the Lotus-Effect. The reflectors are glass lenses with an aluminium reflector behind them, set in a rubber matrix. Whenever a car tyre passes over one of the studs, it pushes it down, past a self-cleaning 'eyelid', wiping water or dirt away. The reflectors are protected by the rubber which is the only part that comes into contact with vehicle wheels.

Catseyes are about to undergo a quantum leap in the form of the 'Intelligent Roadstud'. Instead of a passive glass lens, this uses a light-emitting diode which charges up by day and glows at night. It is visible at 10 times the distance of conventional Catseyes.

Like the hook-and-loop fastener, the Catseye was a one-off that led to no research into further natural mechanisms to exploit. The idea of a systematic combing of nature for useful devices was over 50

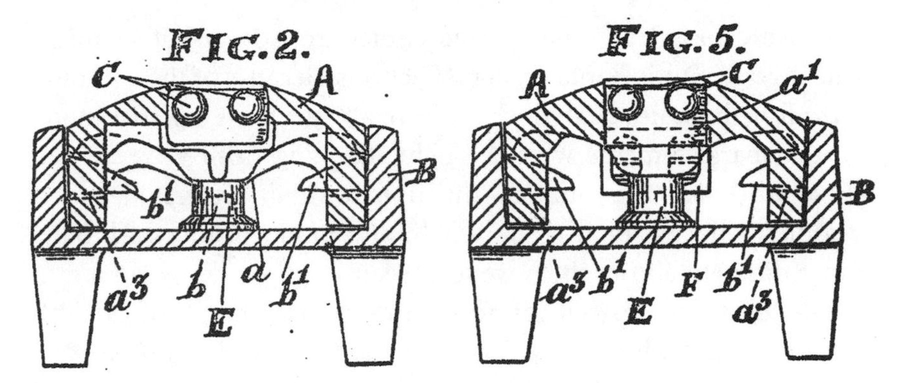

Fig. 5.8 Catseyes® from Percy Shaw's 1934 patent. (British Patent 457,536)

years away and, at the time, most of nature's devices lay hidden in a fog of ignorance: their structure was too small and their chemistry just too complicated.

As opposed to cat's eyes, *anti*-reflection is the purpose behind nanostructured arrays found on the eyes of many creatures: flies and many butterflies and moths have a finely structured array of tiny bumps across the cornea. This is a means of ensuring that most of the light falling on the eye reaches the receptors at the back. The arrays are about 200 nm in height and diameter across the base, tapering almost to a point. In cross section they resemble nipples, hence their name: 'nipple arrays'. Reflection is reduced because it is the transition from the refractive index of air to the refractive index of the eye that causes reflection. The fine point of the protuberances means that the light's first encounter with the new substance is very gradual and there is no abrupt transition to the new refractive index. And since 200 nm is well below the wavelength of visible light (which starts at 400 nm), every particle of light of whatever wavelength encounters this gradual transition.

This principle is used in technical anti-reflection surfaces. It was proposed as far back as 1956, before the advent of nanotechnology, and such structures were only produced from 1973. They are used in double and triple glazing to prevent the cumulative loss of light from reflection at each surface.

Both reflection and anti-reflection can be used as part of camouflage strategies. Sparkling fish scales are an example of reflective camouflage. The sparkle comes from layers of guanine, which create multiply interfering reflections. Because of the contours of the fish, the scales catch the light from all directions and in a sunny sea the sparkles merge with the light dappling from wave splash. These work for fish only in the upper, sunlit waters; deep-sea creatures, whose habitat is sparsely lit and monochrome dark blue, have a different approach, as we have seen.

The ultimate hi-tech disappearing trick is that of Stealth technology and while many details of the US F117 Stealth bomber remain secret, some of the principles are in the public domain and have counterparts in the natural world. Camouflage, as we usually understand it, involves only the visible spectrum, but this is only a tiny portion of the electromagnetic spectrum: for aeroplanes, the microwave region – radar – is the crucial band.

Radar is able to work well in detecting aircraft because the sun's radiation is weak in the microwave region. If the sun emitted a lot of radar, signals bouncing off a plane would be lost in the background 'noise'. But since there is no background, the Stealth has to avoid reflecting microwaves. The first ploy is obvious. If fish scales can catch a gleam whatever direction they are facing, because there is always at least one scale at right angles to an incident ray, so a curved aeroplane is always going to catch a glint of radar. But the Stealth has no curves and so very few possible reflecting angles are presented at any one time. The F117 is reputed to have a radar signature the size of a bumblebee. It is metal that reflects radar and besides presenting few reflective surfaces at most angles, the surface coating of the Stealth allows little radar to reach the metal parts.

In the struggle for life, concealment is just as important as illumination. To systematically evade the sight of predators is an effective life strategy for many creatures – the art of camouflage was one of the earliest examples of bio-inspiration and this brings an intriguing twist to the story. Biomimics are trying to copy, or at least to adapt, nature's structures and devices. Seen from a deep biological perspective, you could say that one earthly life form (*Homo sapiens*) is trying to copy other earthly life forms (spiders, geckos, butterflies, etc). But earthly life forms have been copying each other and pretending to be what they are not for hundreds of millions of years.

The sophistication of nature's visual systems and the life-or-death importance of appearances in predator–prey relationships have created a fantastical biological playground of mimicry. The Russian novelist Vladimir Nabokov wrote in his memoir *Speak, Memory*:

> The mysteries of mimicry had a special attraction for me. Its phenomena showed an artistic perfection usually associated with man-wrought things. Consider the imitation of oozing poison by bubblelike macules on a wing (complete with pseudo-refraction) or by glossy yellow knobs on a chrysalis ('don't eat me – I have already been squashed, sampled and rejected'). Consider the tricks of an acrobatic caterpillar (of the Lobster Moth) which in infancy looks like bird's dung, but after molting develops scrabbly hymenopteroid appendages and baroque characteristics, allowing the extraordinary fellow to play two parts at once (like the actor in Oriental shows who becomes a pair of intertwisted wrestlers): that of a writhing larva and that of a big ant seemingly harrowing it. When a certain moth resembles a certain wasp in shape and color, it also walks and moves its antennae in a waspish, unmothlike manner. When a butterfly has to look like a leaf, not only are all the details of a leaf beautifully rendered but markings mimicking grub-bored holes are generously thrown in.

The principle of camouflage is well known but in nature this is not just a matter of escaping attention. Some creatures disguise the way they look in order to *attract* attention – this is seen in the luring behaviour of bee orchids (*Orchis apifera*), for example, or the orange and green cicadas (*Ityraea gregorii*) that collectively colonize twigs, taking up stations so that the whole resembles a flower spike; any insect that comes expecting to find nectar for honey is consumed. Nature is a seething hotbed of mimicry, aggressive displays, and luring behaviour.

Despite the deep biological importance of appearances, there is a strong strand of *human* prejudice against surface appearances: hence the phrases 'merely superficial', 'on the surface', 'you can't judge a book by its cover'. But nature does not share this prejudice; the whole system of flowering-plant pollination is both superficial and fundamental. And if one creature can out-compete another by being bigger, faster, stronger, better armed, having better senses, another can give the *appearance* of being bigger, stronger etc. Or it can give the appearance of not being there at all.

In his book *Adaptive Coloration in Nature*, the biologist HB Cott brilliantly made the case back in 1940 for the deep biological importance of superficiality:

> The resemblance is literally superficial rather than structural – that is to say, it is generally due to the most ingenious and deceptive disruptive patterns, which give the optical impression of irregular processes and deep interstices – even when painted, as they often are, on the flat canvas on the void abdomen of a spider ... Everywhere we see the same story: the superficial nature of the appearance; and the independent manner of its production.

In battle, it can be just as important to give a false superficial impression as it can be to have a big gun. The word camouflage instantly suggests crude earth-patterned markings on military vehicles and buildings but researchers are now experimenting with subtle systems that adapt to changing conditions. One such new camouflage technique, being developed by Professor Julian Vincent at the Centre for Biomimetics and Natural Technologies, University of Bath, is a bio-inspired camouflage based on the colour- and pattern-generating systems of cephalopods such as squid, octopus and cuttlefish. These creatures can change their colour to blend in with their surroundings. They can also imitate patterns to some extent, as can flat fish such as plaice. Strangely, although the colours of the cephalopods are quite rich, and their colour system is rather like our electronic systems, with three kinds of pigmented cells, which combine in a pointillist fashion to create mixed colours, all the evidence suggests that cephalopods are colour blind. They sense intensity and the colour cells respond to that. But why should nature have evolved an advanced three-colour system in these creatures but no sensory system with which to be aware of it? Julian Vincent says that it is because the cephalopods' colour system is not designed to be seen by other cephalopods but by – or rather not by – their predators!

Cephalopods have a highly developed brain and nervous system and can control the patterns consciously; this could be mimicked in a hi-tech fashion with a large flexible screen that would display computer-generated patterns, but this would be inappropriate technology for something that needs to be very large scale, flexible, robust and cheap.

Nevertheless, in order to prove the concept, just such an elaborate mechanism *has* been developed, by Susumu Tachi, a professor of physics and computer science at Tokyo University. Human beings have fantasized in science fiction about the Invisible Man and clearly it would be a perfect strategy for any living creature to be visible to its own species but invisible to its enemies. In Tachi's experiment, a person wears a hooded cloak displaying on the front images taken with a camera behind the subject, so that whatever is behind forms an image on the front of the cloak. The result is that the cloak blends into the background, blurring the fact that this is a person. Only the disembodied face stands out.

At present, the image is projected onto the cloak from a front projector but theoretically the image could be created by a chip within the garment itself. And if a garment were made of electrochromic pixels, a digital image would form on the cloak. Electrochromic systems are usually made from plastic electronic components, so would be compatible with flexible garments. This would be an enormously elaborate way of disappearing but just to prove the concept it is likely that someone will try it.

If implemented as a military camouflage, such a cumbersome system could make a hundred camouflaged soldiers as expensive as a Eurofighter plane. Julian Vincent's scheme is more modest: the colour changes occur in a gel that can respond by reflecting ambient light as do certain cells in the cephalopods. A certain amount of sensing and electronics is necessary but the aim is to be as naturally responsive as possible.

The target is a form of camouflaged sheeting for military vehicles. Vincent's group is already thinking about the next stage, deploying the camouflage, and that too will involve biomimicry. Instead of manually hauling camouflaged sheeting over the vehicles, they will have a light, sprung pulley system for rapid erection. This deployment mode is based on nature's mechanism for unfurling leaves from the bud in plants such as hornbeam and beech. Such new uses for what is, in practice, origami, both natural and human, are explored in Chapter 8. The combination of two pieces of biomimicry, one from the plant kingdom, one from the animal, shows what a powerful paradigm bio-inspiration has become.

The opposite of camouflage is display and there has always been a connection between human visual display and advertising, and the natural version. Creatures that advertise the fact that they are dangerous use lurid red, yellow and black markings and these are the colours used in human technical warning displays. And flowers were the first advertisers, adopting bright colours and voluptuous shapes to attract insects.

Besides the butterflies and sea mice, another creature has excited biologists and physicists alike with its optical wizardry. This is the brittlestar, a relative of the starfishes. Like starfishes, brittlestars have five arms (fig. 5.9). The arms, each of which is about 5–8 cm long, radiate from a medallion-like centre; they are covered in hairs that undulate in the water and pass food back to the creature's mouth. The brittlestar story began with Gordon Hendler's fieldwork in the Caribbean in the late 1970s. Around the coastline of Belize he observed the five species of Caribbean brittlestars, inhabitants of coral reefs. Hendler, now at the Natural History Museum, Los Angeles, noticed that some of these brittlestars could change colour and were, to some extent, sensitive to light. Before his work, the different colours were thought to indicate different species. Hendler showed that the colour change and light sensitivity were related in a highly intriguing way. The brittlestars are vulnerable to predators and spend most of the day hiding. During the day they are a dull brown colour (although their hairs are faintly green-blue iridescent), but at night when they feed they take on a striking banded appearance: off-white and grey rather than brown. In 1984, Hendler showed that the colour change was due to pigmented cells that migrated.

Hendler found a highly intriguing structure in the light-sensitive species. On the central plate that links the arms was a micro-array of bumps. Unlike the rest of the skeleton, these bumps were

Fig. 5.9 Brittlestars in day (a) and night (b) colorations. Their lens arrays are found on the plate that joins the arms.

transparent. Hendler suspected that this was a lens system responsible for the light sensitivity. The connection between the colour change and light sensitivity was that during the day the colour cells covered the lenses forming 'sunglasses' or lens caps and it was this that produced the brown coloration. At night, the colour cells retracted and the lenses were exposed. One species, *Ophiocoma wendtii*, was clearly more optically advanced than the others. In 1987, Hendler investigated the fine structure of the lens system more closely and claimed that it was a photoreceptor system with nerve cells at the focal points of the lenses.

This work had no impact on the world of physical optics until Joanna Aizenberg, a researcher with a strong array of bio-inspired talents, including microfabrication of crystal structures, brought modern techniques to bear on the question. If such structures are going to result in technical products, equivalents for the natural prototypes need to be synthesized. This requires very different skills to the analysis of the natural structures. Joanna Aizenberg has assembled these skills during a career which has taken her from Moscow University to Israel, where she gained her Ph.D, to Harvard, and finally to Bell Laboratories, New Jersey, the legendary industrial laboratory with an awe-inspiring range of inventions to its credit, including the transistor (1948). And this is where Eli Yablonovitch set optics off on a new course in 1987 with the prediction of photonic crystals and their light-bending properties. Bell is a multi-disciplinary paradise, with expertise on tap in every possible technology.

I went out to Bell Labs from New York on a cold February day with parcels of snow still on the ground. The trip on the New Jersey Transit through the derelict rust-belt around the city of Newark, is a stark reminder of the need for the new technologies bio-inspiration will bring. When you reach Bell Labs you are in need of light and you get it. The foyer display of Bell's expertise, especially in fibre optics, is dazzling. As is the sponge *Euplectella aspergillum* that Joanna shows me (fig. 5.10). Sponge is not a particularly apt name for this creature which builds a hard silica lattice for itself in a pattern reminiscent of Norman Foster's Swiss Re Tower in London (or should that be the other way round – *see* Chapter 9).

Euplectella is an astonishing creature. Joanna says that little is known about its formation; it lives in the deep and is always found fully formed. As an expert in biomineralization – the science of how such mineral lattices are formed, both in nature and human synthesis – she confesses to being baffled by the question of how *Euplectella* grows the lattice. This ornate mineral structure serves a

Fig. 5.10 Joanna Aizenberg with the Venus flower basket sponge (*Euplectella aspergillum*), a sea sponge with optical fibres of superior quality.

simple creature. Sponges are very primitive, perhaps amongst the first multicellular organisms to evolve. They have no organs: they pump water through their tissues and extract from it what food they can. In some ways, they are closer to being a colony of cells than a true organism. But there is no doubting the integrity of the mineral lattice that protects the cells. The basketwork lattice looks like a bouquet from the gods, a harvest festival cornucopia, hence the popular name 'Venus flower basket'. To cap it all, it sometimes contains a pair of mating shrimps and for this reason is often presented as a wedding gift.

But it is not for its beauty and folklore that Bell Laboratories are interested in *Euplectella*, although for Joanna it is part of its charm. The base of the structure is encircled with fine spines or whiskers and these, Aizenberg discovered in 2003, are optical fibres. They are made from non-crystalline silica, as are man-made fibre-optic cables, but, crucially, in the centre of the fibre is a filament of protein. This protein filament confers a huge advantage over man-made optical fibres. Optical fibres are brittle and the organic filament in the *Euplectella* whiskers provides the classic anti-brittle factor, an elastic substance embedded in the structure through which cracks cannot spread (this is the principle of composite materials observed in spider silk, *see* page 68). It is early days, but it is possible that a better optical fibre can be developed using this as a model.

Optical tests on the whiskers showed that they do indeed perform as perfect fibre-optic cables. Exactly what *Euplectella* do with them is not known but it is likely that they harvest what little light there is on the sea floor and signal to other members of the species, as does *Aphrodita.**

Aizenberg's work on the brittlestar is much further down the track. Perhaps the most amazing thing about the brittlestar arrays is that they are made from calcite, a crystalline form of calcium carbonate that has some drawbacks as an optical material. From certain angles, calcite produces double images: it is not a substance that would be first choice for a lens. But the brittlestar lens is orientated along the optical axis, the one direction that avoids the double image.

The brittlestar lens has another trick, linking an ancient creature with 17th-century physics and 20th-century biology. A lens that is only curved on the outside will not bring all the light to a focus at the same point. To correct for this a reverse curve on the underside of the lens was proposed back in the 17th century by the pioneers of modern optics, René Descartes (1596–1650) and Christian Huygens (1629–95). When Aizenberg looked at the brittlestar lens it matched the Huygens/Descartes curve exactly (fig. 5.11).

Thanks to this device, the lens brings light to a focus at a point where there is a nerve receptor, 4–7 micrometres below the lens surface. Evolution has thus achieved a miracle of fine-tuned

* These fibres sound very similar to the fibres of the sea mouse but the *Euplectella* fibres are not photonic crystals. There are no hollow channels running down the fibre, only the protein filament. They guide light in the same way that standard synthetic fibre-optic cables do and they do not produce iridescence.

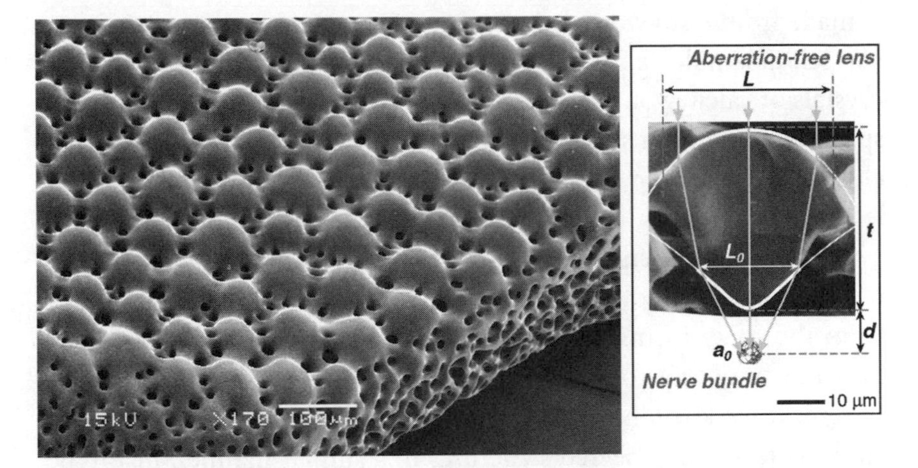

Fig. 5.11 The lens array of the brittlestar is made from a single crystal of calcium carbonate. Each lens is engineered to bring the light to a focus according to the principle first worked out, in the 17th century, by Descartes and Huygens.

---

engineering. The brittlestar lens system is an intricate structure that we should like to be able to grow in the way that nature does. Aizenberg's work on creating technical equivalents for it is described in the next chapter.

Apart from its possible use in the creation of a new type of lens array, the brittlestar is a good piece of evidence in an old quarrel regarding evolution. What use is a half-evolved eye? is an argument often used by anti-evolutionists. Well, the ability to sense any light at all is clearly advantageous for survival and the belated discovery of the brittlestar suggested that no one ever found a primitive eye because they had no idea what a primitive eye might look like.* Now we know: it is made of calcite crystal and has no moving parts but it does a job of sorts. In the country of the blind, the brittlestar would be king.

The most striking thing about the brittlestar lens, and this goes to the nub of bio-inspiration, is that this precision piece of engineering

* Calcite lenses similar to those of the brittlestar have also been found in fossil trilobites from the beginning of vision on Earth, in the Cambrian era, 543 million years ago.

is made from *a single crystal* of calcite. Nature has managed to grow (or self-assemble) a crystal that looks completely different to the crystals of calcite you will find in the geological museums. These are spiky agglomerations in a variety of forms – large double crystals, clusters – but never anything that looks remotely like a brittlestar lens.

So the calcite lenses of the brittlestars must embody one of nature's nanotechnology secrets: how to bend a crystal of calcite from the spiky forms it would have in the absence of living tissue into the smoothly-rounded, exquisitely-patterned and -tuned lens system. If we knew how nature did it perhaps we would be able to 'grow' complex engineered structures in a similar manner. Just to be able to duplicate the brittlestar's feat and grow a lens system would be a breakthrough. Such lenses already have technical uses. But beyond that, perhaps we could grow other miniature components such as silicon chips, instead of etching them into shape?

What is it in the growing brittlestar that allows it to perform such wizardry with a single crystal? We don't know all the details but there is no doubt that crystallization in living creatures is modulated by proteins that produce dramatic changes in the crystal growth patterns. The goal for bio-inspirationists is to understand these modulating factors and to produce crystal structures similar to those of nature. The next chapter takes up the story of the synthesis of nature's functional structures by such organic templating.

Before we delve into how materials scientists are attempting to replicate nature's synthetic feats, you might be still wondering about the passage from Lucretius that forms an epigraph to this chapter: Did he really understand the principle of the peacock's tail, as he seemed to? Was he really 2,000 years ahead of his time? There have been various studies of the peacock over the years but the most detailed appeared as recently as 2003 when a group of Chinese researchers found a two-dimensional photonic crystal in the feathers. To get this in perspective: the sea mouse is one-dimensional – light travels one-way down the tube; the array in *Parides* is three-dimensional, blocking light of some wavelengths in all directions. One method of checking whether it really is the micro-array that is causing the effect is to apply Newton's test: the immersion of peacock feathers in glycerin, which has a very different refractive index to air, abolishes the iridescence, but it returns when the glycerin is removed.

There are some figures in history who really ought to be allowed to come back to see what happened to the subject in which they were such visionaries. Lucretius would be delighted to learn how shape really does cause colour and how things actually look in this minute world towards which his imagination reached out.

CHAPTER SIX

The Molecular Erector Set

Never suppose the atoms had a plan,
Nor with a wise intelligence imposed
An order on themselves, nor in some pact
Agreed what movements each should generate.

No, it was all fortuitous; for years,
For centuries, for eons, all those motes
In infinite varieties of ways
Have always moved, since infinite time began,
Are driven by collisions, are borne on
By their own weight, in every kind of way
Meet and combine, try every possible,
Every conceivable pattern, till at length
Experiment culminates in that array
Which makes great things begin ...

LUCRETIUS, *De Rerum Natura*

The ancient Chinese had tips at the back of their houses where they deposited earth from farming and building activities. The household sewage was also dumped there. In the course of time, the Chinese realized that the material from mature tips of this kind was the best building substance they could find. And the reason is the key to a revolution in materials science that is taking place now. Mehmet Sarikaya, Professor of Materials Science and Engineering at the

University of Washington, Seattle, and one of bio-inspiration's cheerleaders, likes to cite this story as the primal origin of the bio-inspired materials science that is going to transform the way we make things.

The crudity of the Chinese 'experiment' couldn't be further from the intricate, molecularly precise work of modern materials chemists but it does contain the essence of the processes behind the revolution. What was happening in the Chinese tips was that *mineral* matter from the soil, especially clays, was meeting *organic* wastes from the sewage. Proteins in living and once-living tissues have remarkable powers to organize the structure of minerals. Our bones and teeth and the shells of snails, shellfish and other marine creatures are simple minerals – usually calcium carbonate, phosphate or silicon dioxide – structured, in ways chemists didn't remotely understand until recently, by the influence of proteins that control the way they assemble. This process is called 'biomineralization' and the result is a composite many times tougher than any structure that can be made from the mineral alone – minerals are naturally hard but brittle. Even in the primitive conditions of those festering Chinese tips, proteins have the ability to organize clay particles. The result, when mixed to a paste with water, and shaped and dried, is that the clay particles are glued together to form a composite.*

And biomineralization does not just apply to minerals like calcium carbonate. Helen Ghiradella drew our attention to the virtuosity of the butterflies' engineering using the substance chitin, which is a relative of cellulose and starch. Again, it is proteins, and behind them DNA, that template these structures.

* The affinity between proteins and clays goes very deep – it may even be the secret of the origin of life on Earth. Graham Cairns-Smith argued for this in *Seven Clues to the Origin of Life*, suggesting that the self-assembly of minerals, such as the aluminosilicate clays, may have produced some early form of genetic code which then began to organize the structure of organic compounds. Clays self-assemble far more readily than organics, but organics are far better able to carry out specific reactions and to maintain structures and codes. Cairns-Smith argues that the clays led the way until it was possible for the organics to take over. Although it is not part of their programme at all, contemporary materials scientists may be working close to the principles that led to self-replicating (ie, living) matter in the first place.

In the human manufacture of tiny things, the intricate patterns on a silicon chip are first designed on computers at a scale we can see. To simplify an enormously complicated process, what happens next is that the patterns are minimized by lens systems, similar to those in microscopes and known as steppers, and projected photographically onto photosensitive substances; these can then be etched and various layers of materials deposited on top. It is called the top-down process because we first design at our own scale and then get machines to minify these patterns and write them onto silicon. Every year we squeeze more and more tiny patterns onto the same size chip.

There is a curious, completely unscientific, rule of thumb known as Moore's Law. Gordon Moore is the co-founder of Intel, the computer-chip giant, and back in the mid-1960s he proposed the idea that computer chips would double in power about every 18 months. Amazingly, this has held true for over 35 years, but it is now threatened by the physical limits to the miniaturization process. The current Intel chip packs 42 million transistors onto a single chip and the scale of the smallest structures that can be written in this way is 130 nm. The computer industry has an industry-wide standard roadmap that sets the scale for future chip manufacture: for 2007 this is 65 nm.

But when nature set out her roadmap for creation, the top-down route was forbidden. Nature could only begin with the tiny building blocks that were available – atoms – and then start to combine them into ever larger and more complex structures. Atoms are very small and most molecules – which are firmly bonded chemical combinations of atoms – are not much bigger. Even what ranks as a very large molecule, DNA, is still only 2 nm across although, because it has to carry the genetic information, it is very long: in the cell it is tightly packed but strung out end to end a DNA molecule can be millimetres in length (it is about a million times as long as it is wide).

When Richard Feynman said that there was 'plenty of room at the bottom' he meant that he had this in mind: nature starts with such tiny molecules she can build intricate structures with millions of components and still fit them into a tiny space. If nature had evolved something like a Pentium 4 silicon chip, its 42 million transistors could have fitted into a space 3,000 times smaller in area than the technical chip.

In order to utilize some of nature's fabrication techniques, as Bob Full reminds us, we have to understand the mechanisms nature is using. There are two overriding principles. The first is that all nature's structures are under genetic control: DNA is the blueprint, not only for the spider and its silk, but for its web too. DNA lies behind any natural structure we might be interested in. We need to keep this in mind but it is not immediately helpful because even knowing DNA's complete code – as we do for more and more creatures – does not tell us how it *uses* the raw proteins it makes to construct functioning organs. As with spider silk, at first it looked as if GM techniques would enable us to clone the gene for the structure required and then make lots of it, but it is not quite as simple as that: we can't make spider silk that way, although we came quite close, and we certainly can't make gecko foot bristle arrays, or butterfly photonic crystals that way. All is not lost, though, because there are other ways of using DNA than the classic clone-a-gene-and-implant-it-in-another-creature approach that was tried with spider silk.

We understand very well the *first stage* of how DNA makes structures: it makes proteins. Proteins are long strings of chemically linked amino acids and these are added to the protein chain one at a time according to DNA's genetic code. It may seem a very slow and cumbersome way to make living things – one tiny molecule at a time – but it is happening in millions of cells at the same time; and we know it works: you just have to look around you to see that.

What happens when the complete protein floats free of the assembly process? This is the second big principle of molecular erection: proteins *self-assemble*, turning them from one-dimensional strings into folded three-dimensional structures. They do this completely automatically and, unless there is a gene defect, accurately. Proteins are very long molecules and they are made so that parts of the chain attract each other: these attractive regions pull the protein into shape.

Self-assembly is the ability of some chemicals to fold up into interesting structures by themselves: they just happen to fall out that way. On the face of it, it sounds a bit unlikely and the whole business requires some amplification. The first thing to say about it is that, as far as living things are concerned, self-assembly originates with DNA. DNA directs the synthesis of proteins, and the proteins fold up in the way they have been created to do. But self-assembly can take place *without* DNA. It is a chemical phenomenon that occurs in the mineral world (the natural opals we met in Chapter 5 are a good example) and chemists are also increasingly adept at making it happen in test tubes.

You can perform a simple experiment in chemical self-assembly in the kitchen sink. Just squirt a good helping of detergent into warm water and blow into the solution with a drinking straw. A mass of loosely structured bubbles forms, like a soapy Eden Centre. You may think that this is cheating, that the soap bubbles are not permanent structures. But in nature some organisms, such as the tiny radiolarians of the ocean, use the soap-bubble principle as a template: hard minerals are deposited at the junctions of the bubbles until a permanent structure is formed. Materials scientists are investigating every possible kind of template to achieve nanostructures like those in nature.

Scientists would research this for its own sake, even if there were no applications in view. To find out how nature makes something as convoluted as a *Morpho* wing scale is as profound a quest as the search for quarks or the sequencing of the human genome. But a lot of the work does have applications in view: for materials that are tougher and stretchier, self-cleaning, adhesive, electrically and optically functional on the smallest scale, and other attributes that we cannot yet predict. And, after all, no one had the computer in mind when the transistor was invented.

A good place to start to look for the principles of nature's fabrication skills is in the hard structures she builds, such as the shells of shellfish. The first favourite subject was the red abalone (*Haliotis rufescens*), the characteristic edible shellfish of the Pacific coast of America. This was a research topic that was not purely a disinterested search for a secret of nature. The abalone's shell is both large and exquisitely formed, and – here's where the materials science comes in – exceptionally tough. Apart from trying to understand the principle of shell formation, scientists wanted to be able to make

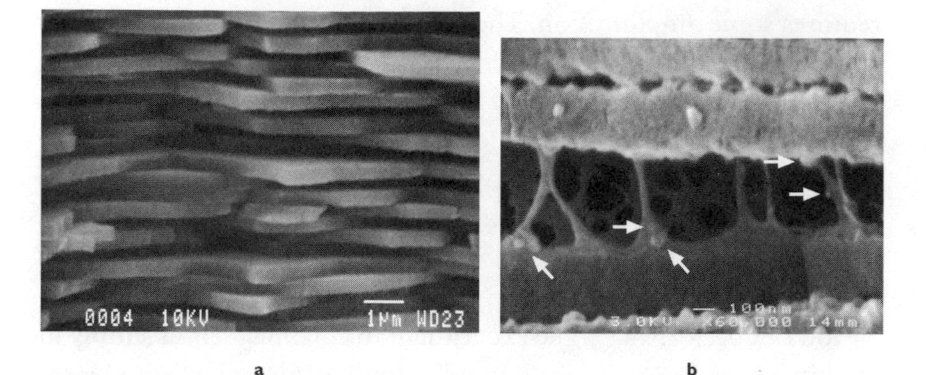

a b

Fig. 6.1 a) The stacked layers of calcium carbonate crystals in abalone, revealed under the microscope. Between each plate of calcium carbonate is a much thinner sheet of protein. b) When the abalone shell is pulled apart the protein stretches dramatically to resist the force.

---

materials as tough as abalone shell, which is made from calcium carbonate – but is 3,000 times tougher than normal forms of the mineral! The inside of the shell, known as nacre, is the tough part and its pearly magenta-to-turquoise iridescence suggests that nanostructures are involved.

Under the microscope, abalone nacre reveals a composite structure with blocks of aragonite (a different crystal form of calcium carbonate to the calcite found in the brittlestar), separated by smaller interleaved layers of protein. There are many unsolved mysteries about abalone but the source of its toughness is not in doubt: it is a composite, designed to stop cracks spreading – under magnification, stressed abalone shows the protein layers stretched by several times their usual size without the matrix splitting open (fig. 6.1). The protein is not only the glue for the blocks of aragonite: it is also the template on which the aragonite blocks are formed. Aragonite will not make blocks like this *without* the abalone protein.

In the early 1990s, abalone became an attractive subject for materials researchers. As so often happens, early progress was surprisingly swift and perhaps led to false optimism. The key research centre was, appropriately, the University of Santa Barbara on the Pacific coast of California, home to the abalone. The first breakthrough was the realization that abalone's mineral synthesis could be induced artificially by slipping a glass slide under the shell of the creature. There is always a suspicion that nature's systems are so complex and delicate that any interference will be fatal. In fact, the opposite is the case: very many natural systems continue to work in the test tube or with artificial irritants such as this glass slide.

The abalone experiment produced a 'flat pearl', a layer deposited on the glass that resembled the characteristic pearly appearance of abalone nacre. The process is thought to involve templating by proteins at the active growing site. The flat pearls show three clear regions, as if a wave of activity has swept over it. At the edge of the layer is a red zone, similar to the outer shell of abalone, then the pearly zone in the middle, then a green zone which is more organic than mineral. The SEM shows that the flat pearl has the same stacked structure as the natural nacre. These pictures of aragonite crystals with their thin interlocking layers of protein are amongst the most suggestive in materials science: they look like multiple piles of pennies on a pub counter. It is not the way we would have thought to build anything but it is so obviously the best way to make a hard, resilient material, the urge to produce a technical equivalent was irresistible. If the abalone protein could do its work on a glass slide, the next stage was obvious. Mehmet Sarikaya takes up the story: 'A lot of people tried to mimic the structure; they said: "Why don't you take out the protein, the protein that actually controls the formation. And you could one day reconstitute the structure."'

Extracting the protein and adding it to solutions of calcium carbonate soon proved that the protein does indeed change the crystal formation of calcium carbonate: specifically, in the presence of the protein, a crystallization process that was producing one form of calcium carbonate – calcite – is switched to produce another – aragonite. But demonstrating the change from calcite to aragonite did not explain how complex shells could be built from this simple mineral. Eventually, the abalone system proved a false dawn. One day, no doubt, it will be fully explained but for Mehmet Sarikaya it was time to move on: 'We spent five years on the problem and we found out that there's not just one protein: there are fifteen proteins and they all act at different times, so they're spatially and temporally

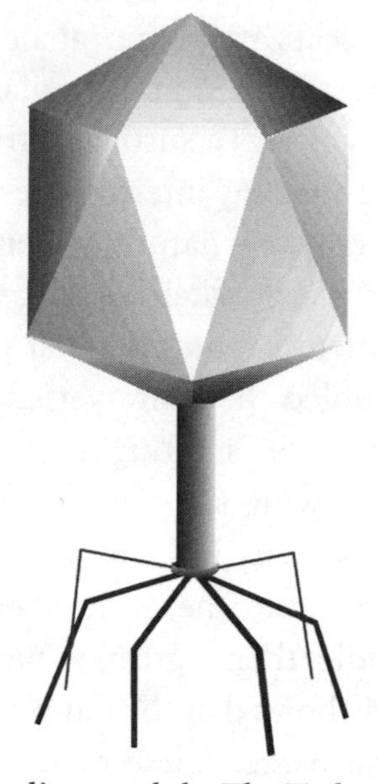

Fig. 6.2 Nature's lunar landing module. The T4 bacteriophage (phage) is an exquisitely tooled nanomachine about 70 nanometres across. Its capacity for self-assembly and its relative genetic simplicity make it an attractive model for bio-inspired materials science.

different. And we're not going to spend thirty years, the rest of our careers, to get to know this; so that's why we abandoned that research.'*

When Mehmet Sarikaya despaired of fully understanding the abalone system, he realized that if we wanted to make useful technical structures, such as bio-inspired nano computer chips, we shouldn't be working with calcium carbonate anyway. We want to

* The practical spur to working with abalone was the goal of super-tough ceramics that could be used for turbine blades, for example. But Paul Calvert, who has specialized in bio-inspired ceramics, says: 'The thing is that abalone shell runs at room temperature whereas the appications you really want in ceramics are things like gas turbine blades and they run at 1,000° plus. I might be able to do that. If you said make me a turbine blade, it's going to run at such and such a temperature and it needs to be tough at that temperature. I could possibly do that. But then you say, I want it to cool down occasionally, and then it gets very difficult.' The problem is the protein component of abalone which, although it confers exceptional toughness, cannot withstand high temperatures.

build structures in silicon, germanium, gallium – the materials of computer-chip technology. But could proteins template such materials? In nature they never encountered them, so what hope was there that any protein could recognize such substances? But astonishingly, they can. It is possible to synthesize new proteins in the laboratory that can recognize and preferentially stick to computer-chip materials. The process is an example of rapid, controlled evolution in the test tube.

The key to the technique is one of nature's most mechanical creatures: the bacteriophage (or 'phage'). When I first came across phages back in the 1960s, their appearance immediately suggested that they ought to have some engineering function (fig. 6.2). Many people have read in *Scientific American* about phages and their amazing properties and in his poem 'The Maverick', David Holbrook articulated the amazement these structures inspire:

> But then, most dreadful of all, the phage,
> Congregating with its sock-tubes on the wall,
> Little more than a molecule, but headed
> Strangely like a man-made gas-tap, like aerial spikes,
> Multiplying itself inside stuff of other creatures,
> Yet with no brain, no heart, no nervous system,
> No soul, no complex flow of organic life, no breath,
> Simply a nut and a spring, and strings of atoms ...

Bacteriophages are some of the strangest creatures on Earth. I say 'creatures' because they reproduce and they have DNA, but half the time they don't seem to be alive at all. Even if you have never heard of phages, you are associated with them intimately: they are very small – about 70 nm across the head and 200 nm long – and they live in your gut. Phages are parasites but we are not the prey. Phages prey on bacteria on the principle enunciated by Jonathan Swift:

> ... a flea
> Hath smaller fleas that on him prey;
> And these have smaller fleas to bite 'em,
> And so proceed *ad infinitum.*

Phages hijack their chosen bacteria, force their way inside, take over their genetic machinery, and make each one produce hundreds of new phages. They do this by landing feet first on the coat of the bacterium and injecting their DNA through the central spindle. The bacterium then dies, splits and the new phages spew out to infect further bacteria. The phages that live in our gut are parasites on the *Eschericia coli* bacterium.

So far, they sound very much alive. But take them out of their bacterial environment, clean them up and bottle them, and they seem to be mechanical crystals; they can sit on the shelf for years, absolutely inert. Phages are so simple, almost their entire chemical structure is known: they are just little geometrical assemblies of DNA and a few proteins. There is no fatty tissue, no specialized organs, no blood or any kind of fluid. Indeed, they have been called 'naked genes'.

Because they are so simple, phages demonstrate some basic biological principles very clearly. One reason that materials scientists have always believed in the possibility of self-assembly is that phages do it all the time. If you break up phages in a blender, they reassemble (fig. 6.3). They do not seem to need any template (as some biological assembly tasks do): the right bits just meet up in the swirling mass of molecules and connect. It is as if you could smash up a mobile phone, pop it into the right solution and, hey presto, it's ring-tone time again! Because, chemically, these things just fall out that way or, to paraphrase Lucretius, the atoms have tried every conceivable combination in the course of evolution but now they have found the right form for the phage they have no choice but to follow their chemical affinities and link up that way.

A new phage is born in the bacterium its parent has colonized. Once the proteins have been synthesized by DNA, they self-assemble into body parts: the 20-sided (icosahedral) geometric figure of the head, the sheath, the legs, and so on. And, finally all the bits stick together to make a complete phage.

We tend to be surprised that nature can produce such perfect geometrical shapes because at the scale at which we customarily view nature geometrical regularity is rare. On a human scale, nature appears imperfect: the two sides of the human face are not identical; trees branch in a haphazard fashion; apples are not spheres. But the

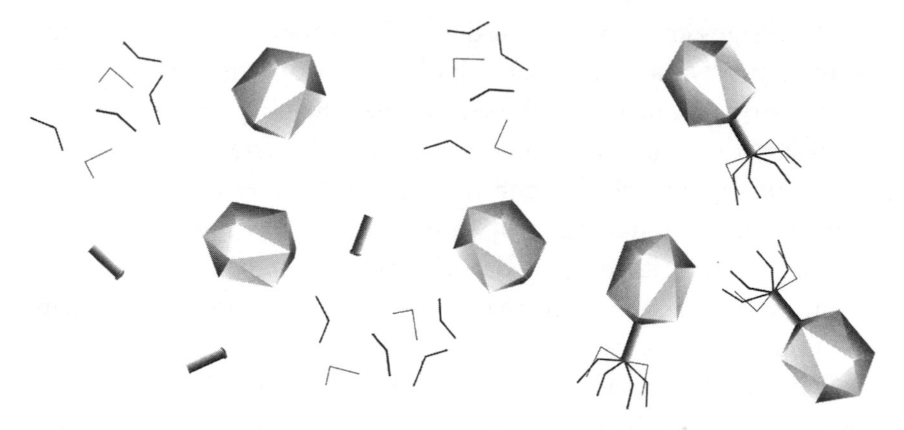

Fig. 6.3 Phages broken up in the test tube can reassemble spontaneously. It is feats like this that give materials scientists the confidence to propose self-assembly as a route to making devices such as computer chips.

closer we get down to the molecules, which are precisely geometrical, the more geometrical order is seen to be present. The icosahedron is one of nature's favourite structures: chemists have been able to make even simple substances self-assemble into icosahedral structures without any DNA templating and without using proteins. Some chemicals just fall out that way because the icosahedron is an efficient, least energy way of filling space.

The phage is produced on a molecular assembly line and just as the parts of a car have to be assembled in the right order, so they do with phages. This is achieved in the smartest way possible. When two parts come together the structure changes slightly and this makes it receptive to the next stage, and so on. So the parts of a phage will not stick together *until the necessary preceding stage has occurred.* This is why it is possible to reassemble a phage: the bits will not stick together out of order.

All this is very good news for materials scientists. Nature's way of making proteins is well understood and can be adapted to make technical proteins, and if these are structured correctly they will self-assemble automatically. The joy of bio-inspiration is that we know these things work perfectly, because nature already does them. Until now, most inventors have never really been sure their gadget would work until it did.

The phage can help us to produce self-assembled technical structures in two ways: one uses the biological properties of the phage itself to organize materials, the second uses the phage as a metaphor. To look at the second method first, George Whitesides at Harvard, a master of chemical self-assembly, has taken the idea of the smashed mobile phone rebuilding itself in a broth almost literally. He has shown that quite large polyhedral electrical components with solder connections can wire themselves up by being popped into a warm solution that melts the solder.

Although the device Whitesides and his team made would be easier to make by hand – the components were quite large faceted balls (5 mm octahedra with the corners cut off) – the point of the experiment was to prove the concept on a large scale with a view to scaling it down later. An advantage of this approach is that it is a parallel process: if it works, all the bits more or less link up at the same time. In a normal assembly line, humans or robots have to do things one at a time. On the nanoscale, these link-ups would happen in their millions simultaneously.

But how do you get things to link up in the correct way? Because of the way the components are faceted, there are only a few ways they can nestle. Whitesides' polyhedra have two kinds of faces, hexagons and squares: the hexagons have a tiny light-emitting diode (LED) and the squares have four solder dots which are electrical connectors. All of the square faces have the four connectors so that if any two square faces link up they will make electrical contact.

The balls are placed in the warm solution and gently tumbled for around an hour. The solder dots become receptive to fusing but if they meet an LED face they won't find any solder to fuse with and the shape is wrong. Eventually, all the components connect themselves up correctly. Whitesides assembled 12 units in this manner, connected them to a battery and the LEDs lit up. Although this is just fairy lights, the units could be transistors or other components. The next stage is to devise strategies for much smaller self-assembly, and this Whitesides has started to do. In 2002, using this technique, he created an array of 1,500 silicon cubes on an area of 5 sq cm in less than three minutes.

So far so good. But how can phages help us find those proteins needed to template the structures of computer-chip materials? In nature, proteins can direct the formation of abalone shell or the wing scales of a butterfly – this ability has evolved over hundreds of millions of years. But natural proteins have never encountered the substances used to make electronic devices. Genetic engineering techniques have enabled scientists to test millions of random peptide sequences (peptides are short sequences of amino acids; effectively mini-proteins), in a process known as bio-panning, to see if any have the ability to interact with inorganic computer components. The test peptide sequences are created on the surface of the phage heads.

Out of all the millions of peptide sequences on the phage heads, a few seem to recognize computer chips as long-lost friends and bind to them. The strongest binding proteins can then be sequenced to determine their structure. Once the sequence of a binding protein is known, it can be made in quantity by standard GM techniques.

What an extraordinary thing to happen. Here is one of nature's previously hidden powers: proteins – some proteins – just love to get tangled up with computer chips. It seems they have that capacity: things just fall out that way.

The technique is known as phage display for the creation of Genetically Engineered Proteins for Inorganics (GEPIs). Finding a library of proteins that bind to the inorganics needed to make devices is a start: the next stage is to use the proteins to template the inorganics to make working structures.

In 2001, Angela Belcher discovered a brilliant shortcut by using the phage not just to evolve a protein that binds to an inorganic compound – zinc sulphide in this case – but to create a nanodevice. She took the inspired step of letting the phage itself assemble into a film. Instead of the phage just being the means to obtain the protein that binds to the inorganic, the phage also becomes the fabric holding the zinc sulphide particles in position. Because phages are very small, the zinc sulphide particles end up only 3 nm apart.

Such organic/inorganic hybrids could have many uses. The primary goal of Belcher's work is the quantum-dot computer element. The essence of a computer is a very small switch that can flip between two states – on and off. The ultimate prize, which has already been achieved as a one-off experiment, is the smallest switch that can be conceived: one that can be flipped by a single electron: the quantum dot. Nanoscale arrays such as those devised by Belcher are potential quantum-dot devices.

Angela Belcher began as one of the abalone researchers at the bio-inspiration forcing-house that was University of California, Santa Barbara, in the early 1990s under Dan Morse. She was at the University of Texas when she wrote the phage paper and soon after moved to the Massachusetts Institute of Technology (MIT) as Associate Professor of Materials Science. So successful was Belcher's approach that at a meeting in New York City in July 2003 she predicted that her lab would produce a field-effect transistor by this technique within six months (a field-effect transistor is the basic element of a computer chip). This was obviously a little optimistic because it has not happened yet but what is so special about producing such a thing?

The answer lies in the size of the zinc sulphide dots – 3 nm in diameter, compared to the 130 nm of present-day silicon technology. If successful, the phage technique will take us past the limiting barrier of lithographic chip manufacture. Those nanodots could be the way to extend the life of Moore's Law.

Belcher's technique is not the only way of using phages. Once the proteins that bind to an inorganic substance have been identified, their affinities and self-assembling properties can be used to create structures without the phage. The technique is still in its infancy and we are a long way from making anything as intricate as a *Morpho* wing scale. Belcher's shortcut was a stroke of genius in cutting through what is going to be a long process.

In their natural habitat on bacteria, phages can spring into life, injecting their DNA down through the central spindle into the bacterium. This requires energy and, at this point in its life cycle, the phage is behaving as a motor. In fact, nature abounds in miniature engines. There are tiny engines inside every living cell and these engines, the powerhouses of life, are virtually identical in all living things, from bacteria to ourselves. The universality of the way that cells use sugars and carbohydrates to produce the energy for life processes has been known since the mid-20th century – it was one of the great triumphs of biochemistry, but the revolving motor

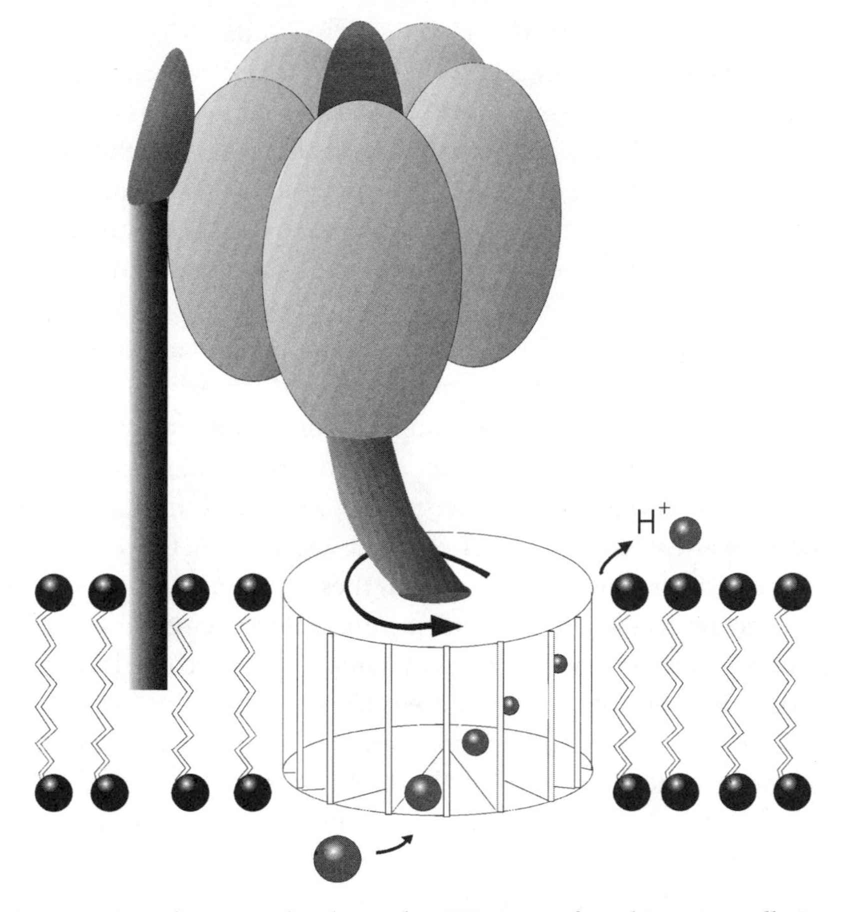

Fig. 6.4 Nature's nanotechnology: the ATP motor found in every cell. One of the great surprises of modern biology has been the revelation of this miniature electric motor at the heart of all living things.

structure at the heart of this process was among the treasures lost in the Blind Zone of nature's nanotechnology.

Energy is needed to carry out all of life's processes and evolution has come up with a one-size-fits-all molecule for the job. This is ATP (adenosine triphosphate): a three-letter acronym that ought to be as famous as DNA (they are chemically related) – between them they cover a great deal of the action of being and staying alive. ATP is made in a complex piece of protein nanotechnology called ATP synthase: this is an enzyme that speeds up the formation of ATP by billions of times.

It was known that the chemical reactions involved in the breakdown of glucose in cells were cyclical, with electrically charged hydrogen ions being passed down long chains of molecules in a nano-version of pass-the-parcel. What was *not* realized was that these cyclic reactions end up in a real solid structure that actually does rotate! This revolving structure is ATP synthase or the ATP motor. Its structure has now been very accurately determined and it looks satisfyingly mechanical (fig. 6.4). There are two parts connected by a shaft. One end, lodged in the cell membrane, looks like a waterwheel, and it is here that the reactions produced by the breakdown of glucose enter the motor: this end is a proton pump with the protons taking the part of water in a waterwheel. At the other end are six protein structures clustered around the spindle, like an exotic mushroom. These are anchored to the interior of the cell and don't revolve.

One of the fascinating things about these motors is that in bacteria they are reversible. In an electric motor, electricity goes in and motion – rotation – comes out; in a dynamo, the machine is spun by an outside source of energy and electricity comes out. In bacteria, ATP synthase can be either motor or dynamo. If the products of glucose breakdown are applied, the ATP motors revolve and produce ATP. When ATP is provided, the motor turns in the opposite direction, anticlockwise, and pumps protons out of the cell.

The ATP system evolved very early and it has been conserved throughout all the momentous development of life during evolution. The differences between the bacterial and the human ATP motors are slight but revealing. In bacteria, the motors are lodged in the outer cell wall; in higher creatures, such as humans, they are lodged in the outer wall of a part inside the cell called the mitochondrion. Although it cannot be proved, most biologists believe that human mitochondria are the descendants of primitive bacteria that began to live inside other cells for their mutual benefit billions of years ago. Mitochondria even have their own DNA: mitochondrial genes do not go through the human reproductive process – they are passed on directly in the egg cells. This is suggestive evidence for the 'captured bacteria' thesis.

These tiny motors in the cell are hugely important for bio-inspiration because making such things has always been one of the

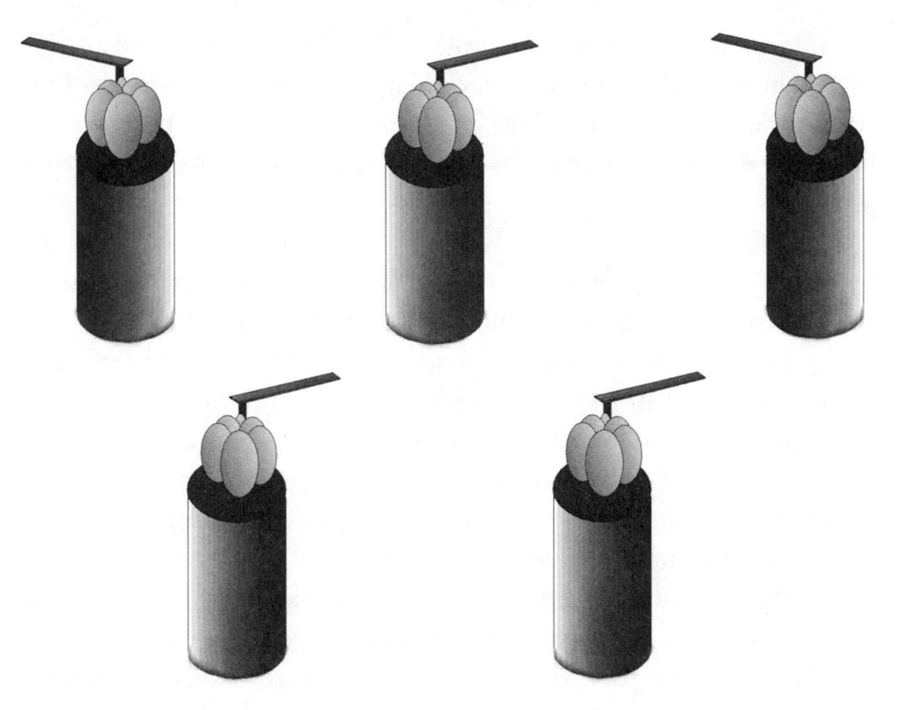

Fig. 6.5 A nano windfarm! Tiny propellers attached by chemical self-assembly to ATP motors mounted on a glass slide create a nano windfarm that combines natural and technical components.

goals of nanotechnology. Seeing is believing, so when the protein motor was discovered scientists looked for ways to make the rotation visible: might it be possible to attach a propeller to the shaft of the motor and observe the rotation through the microscope? How this was achieved is an object lesson in the use of protein recognition technology in bio-inspiration.

ATP motors are robust and can be extracted from cells without damage. To function as a motor outside the cell only half of the molecule is required – the 'waterwheel' can be discarded. You need to genetically engineer the protein motor so that it can be stuck on an inorganic base and you also need to find a way to attach the propeller to the shaft of the motor. Initially, the motors were engineered to stick to nickel posts placed in an array on a glass slide. So when a solution containing the motors is swilled over the posts, they attach themselves by their sticky ends. The result was something like a nano windfarm, with the nano windmills sitting on top of their posts (fig. 6.5). The propellers are then attached by coating them with a protein that sticks to the spindle of the ATP motor. The wind in this case is a solution of ATP which acts like any fuel: it needs to be replenished otherwise the motors stop.

So this method is really Whitesides' technique of cooking up precise engineering components by self-assembly, put to real use on the nanoscale. Nature's nanomotors are 8 nm in diameter and 14 nm long, and they are powerful – a spoonful of ATP would have the rotational power of a large Mercedes engine – and they can power a propeller 1,400 nm long. Once attached, the propellers can be seen to revolve and, wonder of wonders, they are large enough not to need an electron microscope – the waltz of the revolving propellers can be seen through a light microscope.

In the first experiment, in 2000, the self-assembly process was imperfect: only 5 of 400 propellers were observed to rotate – at a rate of 0.7–8.0 times per second. The others were assumed to have misaligned, stuck to the motor case rather than the spindle. Despite the low hit rate for connecting propellers, this was one of the first triumphs of the new hybrid engineering. But this is not the end of it. Although the point of attaching the rotor was simply to be able to see the rotation at work, it was an enormous boost to the goal of making a nanomotor. How could this be done?

In fact, Carlo Montemagno and his team at the University of California, Los Angeles, who did the ATP work, have developed an elegant new way of harnessing biological energy. By using the hybrid biological/technical processes discussed here, instead of using individual ATP motors, they have been able to get rat muscle tissue to grow on fabricated devices that can then contract, powered by nature's energy source, glucose. This is a huge step along the road to functioning hybrid bio-inspired devices. Biological muscle could power some of the micro-machines of the future.

The abalone work was not a complete dead-end: the essence of abalone synthesis lies in finding a template to modulate the crystallization of calcium carbonate. Even if the abalone system is for the moment too complicated, there are other more promising ways. Just as proteins can template technical materials such as silicon, the

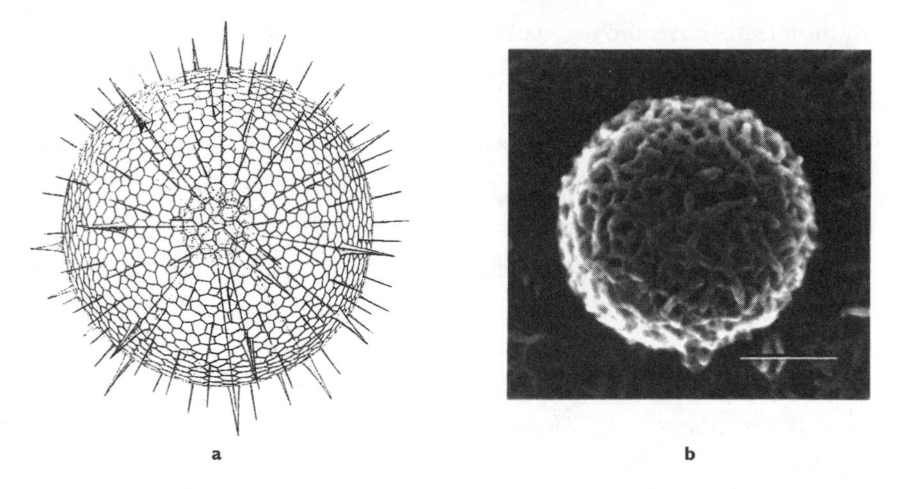

Fig. 6.6 a) The radiolarians of the ocean often display the classic mix of hexagons and pentagons familiar from Buckminster Fuller's geodesic domes and the modern football. This is one of the famous drawings by Ernst Haeckel from the voyage of HMS *Challenger* 1873–6. Such structures are a challenge to materials scientists; b) one of Stephen Mann's radiolarians made by a similar process to the natural radiolarian synthesis.

---

templating of calcium carbonate can be achieved by means other than nature's own proteins. Inspired by the way the brittlestar constructs its optically perfect lens from a single crystal of calcite, Joanna Aizenberg has developed new ways of modulating the crystallization using synthetic templates.

Another way of self-assembling mineral structures involves those soap bubbles (*see* page 139) as a template for creating mineral structures from calcium carbonate and silicon dioxide. Silicon dioxide is a technical material much used in the computer-chip industry but it is also used structurally by nature (especially in the oceans), where it is usually called silica.

Radiolarians are microscopic marine creatures about 0.1–0.25 mm in diameter that build ornate silica basketwork cages for themselves. They became known largely through the work of the German biologist Ernst Haeckel on the voyage of HMS *Challenger* (1873–6). Haeckel's drawings for the official record of the Challenger voyages have fascinated people ever since for their fantastic forms (fig. 6.6), variations on an underlying theme, rather like snowflakes.

Radiolarians have such suggestive geometrical pat-terns that they cry out for an explanation in terms of physical forces. The broad outline of how they are formed was deduced by the great Victorian/early 20th-century pioneer of bio-inspiration D'Arcy Wentworth Thompson in his classic book *On Growth and Form* (1917).

Thompson has the strongest claim to be called the founder of bio-inspiration. He was one of the great Victorian polymaths (although he lived to 1948, his classical scholarship, omnivorous learning and proudly independent stance stamp him as a Victorian). Thompson was a naturalist 24 hours a day. His daughter, in her biography of him, *D'Arcy Wentworth Thompson, the scholar-naturalist, 1860–1948*, recalls:

> On the shore his pockets were filled with shells and sea creatures wet and slimy, and in out-of-the-way places like Roundstone in County Galway, little envelopes were filled with sand, for there the shore is of a brilliant whiteness with sand composed of exquisite microscopic shells from the bed of the Atlantic. And periodicity or number meant so much to him that he was always observing it: counting the petals on a flower, the ripples in the sand, the feathers in a bird's wing, the steps in a church tower.

D'Arcy Thompson looked at structures in nature and showed how their form could be explained by the forces that act on them. For example, the human femur is reinforced in the direction of the lines of stress that act on it. D'Arcy Thompson's emphasis on the shaping influence of the forces of nature was thought to be heretical in terms of Darwinian evolution. Because, of course, in Darwinian evolution nothing that happens to a creature during its lifetime can be passed on to its offspring.

A blacksmith is a good illustrative example. If an averagely built youth becomes a blacksmith he will develop massive, strong forearms. Muscle and bone can respond to persistent extra loads by growing dramatically. But, if once he has acquired his new physique, our blacksmith has a son, the son will have a physique similar to that of the blacksmith's original state, randomly slighter or more powerful, but certainly not disproportionately large in the forearms.

The germ cells are formed when a person is born and they do not change between birth and reproduction.

Why then are dynasties of blacksmiths always apparently born to be blacksmiths, with massive forearms ready made? Because the trade has obviously attracted men who already had the right physique and this genetic physique *can* be transmitted to future generations.

In a similar way, those creatures that are well suited to the forces of nature have a better chance of surviving and thus of reproducing. In this way, organisms perfected by millions of years of evolution look as if they have been fashioned directly by the forces of nature, although in reality the influence has been indirect. A mutation that produces a slight improvement in the adaptability of an organism to the forces of nature will increase its survivability and its chances of reproduction. The brittlestars that developed lenses had an advantage over those that did not.

When the radiolarians are growing they are surrounded by a bubbly froth and Thompson used physical chemistry to explain how froths take up certain forms in order to minimize the surface-tension forces between the bubbles of the film and its surrounding medium. This is the cue for blowing into that bowl of detergent water. When you have got a good head of froth you will see that each bubble is surrounded by three immediate neighbours and that the angles between them are equal. Overall, this creates a hexagonal pattern: in two dimensions the hexagon is always the preferred shape nature uses to fill space because it minimizes materials and surface energy, and maximizes strength. That is why it appears so often in nature: in bee honeycombs and wasp nests, for instance.

But there is a geometrical twist to this. To satisfy the least energy requirement, the bubbles collectively fall into the shape of a large sphere but the laws of geometry forbid a mesh of hexagons to close in to form a circle. This rule was discovered by the 18th-century German mathematician Leonhard Euler (pronounced 'Oiler'). To form a sphere from hexagons, a sprinkling of other-sided figures is needed. Five-sided figures are best (pentagons) although nature is sometimes imprecise and uses the odd square or heptagon (pentagons can clearly be seen in the radiolarian in fig. 6.6).

Hexagonal/pentagonal structures in nature and technology at every scale are explored in Chapter 9. In radiolarians, silica is deposited at the junctions of the frothy bubbles. As this thickens and hardens, it preserves the shape dictated by the equilibrium positions of the bubbles. Radiolarians are bubbles and froth made permanent.

No one followed up Thompson's work for a very long time. His lack of interest in the mainstream biology of the day relegated him to a backwater. For decades, it seems, he was valued chiefly for his prose style: his work was damned as *magnifique mais ce n'est pas la science.* Stephen Mann, a materials chemist who has synthesized artificial radiolarians, says: 'D'Arcy Thompson gave a very physical chemical description and as you know he fell out of favour for a long time because molecular biology and the deterministic view of form was dominant and D'Arcy Thompson was always going on about soap bubbles.'

And if his own colleagues in biology paid him scant attention you could hardly expect chemists to know about him. So Thompson's ideas on the radiolarians had no practical consequences for over 70 years. Finally, in 1995, Geoffrey Ozin and his colleagues at the University of Toronto made a detergent template that could organize the formation of stable spheres of an aluminophosphate clay as large as 1–2 mm in diameter, with cratered and hexagonally meshed surfaces.

At the same time, Stephen Mann, now Director of the Centre for Organized Matter at Bristol University, was also producing radiolarian-like structures by letting mineralization of calcium carbonate occur at the boundaries of micro-emulsions (fig. 6.6). The salad-dressing vinaigrette is a classic example of a micro-emulsion; it is a water-in-oil emulsion, which means that tiny drops of water are suspended in oil (mayonnaise is the reverse: drops of oil suspended in water). In vinaigrette, the stabilizing detergent is provided by the proteins in ground mustard.

Mann experimented with various ratios of oil and water, resulting in different structures and pore sizes. In fact, in order to create spherical radiolarian-like structures Mann cheated a little. His reactions did not produce spheres entirely by self-assembly: they formed on polystyrene bead templates, which were then dissolved away, leaving the mineral shell. But the reaction was radiolarian-like in turning an emulsion foam into a hard calcium carbonate network and Mann was the first materials chemist making artificial radiolarians to cite D'Arcy Thompson in his papers.

What the artificial radiolarians showed, and it has been confirmed by countless experiments since, is that quite simple chemical reactions can produce the kind of complex structures we see in nature; with the more subtle templating of proteins or technical equivalents for them, it is hoped, eventually, to match nature's intricate architecture.

The delicate structures made by Mann and Ozin were beautiful artefacts but they had no obvious use. Indeed, Mann is not sure how useful their characteristic features are to the creatures themselves. Of the coccoliths, a particularly fine piece of nanoengineering in which interlocking plates are sculpted, like Joanna Aizenberg's brittlestar lens, from a single crystal of calcium carbonate, he says: 'They are beautiful but I don't think there's any biological function associated with that; I don't think these coccoliths think: "Gee, what a great guy over there."' But human beings do think that and Mann's micrographs of his artificial radiolarians won the Vinci of Excellence Trophy in the LVMH Met Hennessy/Louis Vuitton Science in Art Competition in 1996.

Mann and Ozin's work showed how mineral structures could be formed from a physical foam template rather than a biological one. Taken together with Aizenberg's pseudo-biological templating, the phage display technique and the inverse opals we met in Chapter 5, it is clear that there is no shortage of shaping principles to hand for the bio-inspired fabricator.

There are many more techniques in current use in materials science than can be accommodated in this chapter. In fact, there is an explosion in the subject and the barriers between the physical, chemical and biological approaches to materials are rapidly breaking down. The stakes are high: nanostructured electronics and photonics will be the next epochal stage after transistors (1947) and the microprocessor (1971). Michio Kaku, Professor of Theoretical Physics at the City University, New York, goes so far as to claim that this revolution is necessary to avoid a world economic recession caused by the collapse of Moore's Law. But you do not have to accept this to realize that the breakthrough into a size realm 100–1,000 times smaller than what is possible now would create remarkable possibilities.

That is the positive aspect. The potential negatives of harnessing biology and nanotechnology were signalled when Eric Drexler, the John the Baptist of nanotechnology, created the 'grey goo' myth. In his influential *The Engines of Creation* (1986), he suggested that nanorobots, created in atom-by-atom assemblers, would be able to reproduce, escape from human control and quickly spread like a plague to cover the world in grey goo. When we assess the modest achievements (as opposed to the promise) of bio-inspiration even 20 years later, this was pushing it a bit for 1986. Which Drexler himself now seems to recognize, judging by his more recent statement (*see* page 25).

Drexler's recantation is good news, but the idea has been out there for 20 years, replicating, and may now be beyond the point where it could be reined in. The same idea is at work in Michael Crichton's bestselling thriller *Prey* (2002). In *Prey*, the work of Sarikaya, Belcher and Aizenberg has already come to fruition.

In a fabrication plant in the Nevada desert the Xymos Corporation is manufacturing nanobots – tiny self-contained machines using a hybrid of nanotechnology and genetic engineering: 'There wasn't much difference between creating a new bacteria [*sic*] to spew out, say, insulin molecules, and creating a man-made micromechanical assembler to spit out new molecules.' The ostensible purpose of the Xymos fabricators was a robot mini-camera that would be able to travel through the blood vessels. This is a real goal for some nanotechnologists, but for the sake of a good thriller, Crichton endowed his nanobots with the property of replication, the ability to learn and to evolve dramatically sophisticated behaviour in a matter of days.

Although the plot of *Prey* is outlandish, Crichton had read a lot of scientific papers (which he cites at the end) and some of the science rings true. He understood why Drexler's programme was not realizable. Drexler envisaged the assembly of nanostructures atom by atom, rather like an automated car factory on a minuscule scale. But atoms won't stay put to allow you to assemble them. On this level, everything is a boiling cauldron. Nature has managed to harness this energy to create nanostructures that can survive this constant buffeting. More than that, she goes with the flow and actually uses the propulsive power and the random motion of atoms as the means to assemble them correctly.

It is strange that the goal of self-replicating autonomous objects is taken to be the goal of bio-inspiration. There *is* a technology that could create new organisms if we chose to go down that road: it is called genetic engineering and it works entirely within biology. The nanobot already exists: we met it earlier in this chapter – it is the bacteriophage, a primitive chemical machine that can reproduce, in its own funny way, inside bacteria, but it is perfectly natural. The goal of materials bio-inspiration is the production of *technical* devices that would be no more alive than the chip in computers today. Whatever else it means, bio-inspiration does not mean the imitation of life in the sense of creating it.

Bio-inspired materials science has attracted some of the most gifted scientists working today. They have created a new body of techniques that blend molecular biology with advanced synthetic chemistry and nanofabrication. The initial inspiration of abalone and other natural composites has given way to genuinely new human fabrication technologies and the substances of life turn out to have a surprising affinity for the technical materials of computing and optics.

All of the bio-inspiration considered so far has been at the nano- and micro-scale but sometimes it is nature's larger structures we would like to emulate, such as the flight mechanism of insects, or the way that nature folds leaves and insect wings, or the way she makes large stressed structures, such as the human frame. All of these structures have their roots in nanostructures because that is the only way that nature can make things on a large scale: by piling nanostructure on nanostructure until the larger scale is attained. In the next chapter, the focus will be on the insect wing itself rather than its micro- and nanostructured surface.

CHAPTER SEVEN

Insects Can't Fly

> The fundamental laws of aero-
> nautics, dynamics, and what ever
> must soon convince the unbeliever
> that bees were built to such a model,
> they scarcely could do more than waddle.
>
> The ratio of their body weight
> to wing-span, he could demonstrate,
> precluded takeoff, much less flight.
>
> SHEENAGH PUGH,
> 'Bumblebees and the Scientific Method'

When Primo Levi asked, 'Would all the philosophers and all the armies of the world be able to construct this little fly?' it would never have occurred to him that any army would *want* to do such a thing. But they do because the fly is a miracle of robotics, turning at right angles on a dust speck, hovering, avoiding obstacles (except the visually deceptive windows) with ease: a wonderfully manoeuvrable little dogfighter with, according to Rafał Żbikowski, an engineer at Cranfield University working on flapping flight, 'less computational power than a toaster'.

Not so long ago, aeronautics engineers were disdainful of natural flight mechanisms. In *The Simple Science of Flight*, Professor Henk Tennekes, an early enthusiast, tells how, back in 1969, he dared to talk about bird and insect flight in his courses. His boss informed him: 'In your class you seem to have talked about geese and swans. I cannot condone that. Our profession – mine, and I trust yours, too – is a branch of engineering. Animals that flap their wings are none of our business. Please restrict yourself to airplane theory.'

Although the Wright brothers studied vultures in flight, aircraft flight owes little to birds or insects, and for a long time conventional theories of flight could not account for the amount of lift generated by insects. The equations said that such creatures should not be able to fly. It is chastening not to know how they do it. If we could fly at twice the speed of sound from London to New York, how is it that we did not understand this alternative form of aerodynamics?

But in patient work over recent decades, biologists in Britain and America have unravelled much of the complexity of insect flight and brought the science to the point where engineers can contemplate making flying vehicles that flap their wings.

But what kind of plane would flap its wings? Clearly, it is not going to be a rival to the 550-seat Airbus A380. The US Defense Advanced Research Projects Agency (Darpa) initiated the research in 1996 with a $35 million development programme to develop 'micro aerial vehicles (MAVs)'. The idea goes back to 1992 when Darpa conducted a workshop entitled 'Future Technology-Driven Revolutions In Military Operations'. One of the topics of that workshop was 'mobile microrobots'. Various studies demonstrated that the concept was feasible, so the call went out. The Darpa project formally ended in 1999; they were, as so often, a little ambitious with the timescale but the work goes on in several laboratories.

The target of the Darpa project was a flying vehicle that could fit into a 15 cm sphere, weigh no more than 140 g, fly for up to 2 hrs and have a range of 10 km, operate in winds of up to 50 kph, and be able to manoeuvre without a remote pilot sitting at a control panel guiding it. This specification does not require or necessarily imply flapping flight, but two teams, led by Ron Fearing at the University of California, Berkeley, who we have already met for his gecko work (*see* page 85), and Rafał Żbikowski at Cranfield University in the UK, have decided to go down the flapping route, and in very different ways: Żbikowski is designing a machine at the 15 cm limit specified

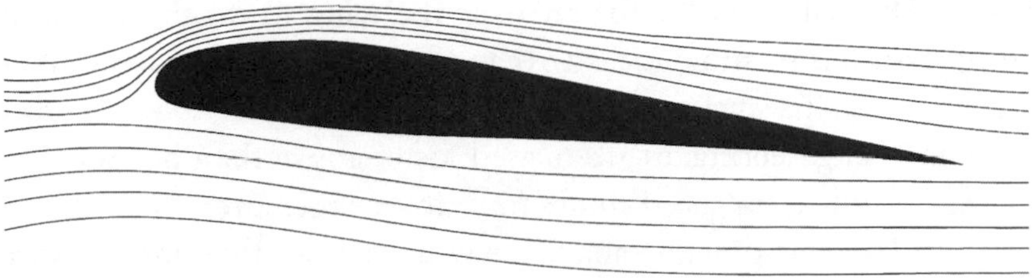

Fig. 7.1 In an aircraft's wing, the air passing over the upper surface moves faster than the air passing underneath; this creates higher pressure beneath the wing and hence lift.

by Darpa, while Fearing is working on the much smaller scale of the blowfly (a large housefly with a 25 mm wing span).

Before looking at how insects fly it is worth going over how aeroplanes do it. Despite the innovations of 100 years of aviation – from the first biplane that hopped off the grass to the Lockheed Blackbird that can fly at 2,000 mph at 85,000 feet – what gets a plane into the air and keeps it there has not changed, except perhaps in the case of the Harrier jumpjet which rises vertically on sheer engine thrust before switching to normal flight.

To fly, wings that are moving horizontally through the air need to generate a powerful lifting force from the air moving past them. To do this, the upper surface of an aircraft wing is more rounded than the lower and the front of the wing more rounded than the rear. So a section through the wing, known as the aerofoil, looks like an elongated teardrop (fig. 7.1). The airflow, hitting the front edge, splits to pass over and under the wing, and produces lift. We all know that it works because we trust our lives with it but the reason it does so is somewhat counter-intuitive.

If one imagines the air seeded with visible particles, then the particles are carried by the air and their pathways are known as streamlines. Before the air hits the wing the flow is symmetrical, so the streamlines are parallel and equally spaced. When the air hits the wing, the airflow splits asymmetrically due to the curve of the wing: over the wing the particles, and hence their streamlines, are forced closer together. This means that the mass of air over the wing has to be squeezed through a tighter area than the corresponding mass below. The only way for the air over the wing to do that without compression is by moving faster than the air below. The streamlines are no longer parallel, but are curved and stay closer together over the wing (ie, the particles are carried faster above than below). The energy for this extra speed comes from the air pressure, which falls as a result. Thus the pressure above the wing is lower than the pressure below and the result is an upward pressure on the wing: lift.

There are fundamental differences between aeroplane and insect flight. For a heavy aeroplane, the main battle is with gravity. But for small creatures, gravity is much less important – insects can fall from a great height and come to no harm. This is because air resistance depends on the surface area presented by a body whereas gravity acts on the three-dimensional mass. As we saw with the gecko, as bodies get bigger their mass becomes disproportionately large compared to their surface area, and vice versa. So small creatures fall through the air slowly: they are, in effect, their own parachutes. All things left to fall through the air attain a terminal velocity at the point at which the air resistance matches gravity: a sphere of 1 m diameter reaches a terminal velocity in air of 1,138 kph, one of 1 mm only 13 kph. But when it comes to flapping flight, if the air breaks the fall of a light creature it will also restrict its flapping motion to some degree.

With insect wings we are in new territory compared to the structures that give us the Lotus-Effect, the gecko's adhesion and spider silk: insect wings are moving parts and they move fast. It is often said that nature and the human engineer have completely different approaches to movement. We use wheels, axles, ball bearings and linkages whilst nature uses stretchy muscles and bendy hinges. A fly's wings beat at up to 200 times a second but they have no bearings. That ferocious vibration is sustained by a hinge made of resilin – this is *the* resilient material. So for the fly, nature had to come up with a hinge that can flex millions of times without breaking. Human engineers find this challenging.

The patterns made in the air by an insect's wings are complex and its degree of control very subtle, but the actual wing movements are restricted. There are three basic movements and these can be combined in varying degrees. Simple flapping is just mechanically up

Fig. 7.2 The wing-beat cycle of a fly. In sweeping forward, reversing the wing's angle, and bringing the wing back to the starting position, the wingtip traces out a rough figure-of-eight pattern.

and down with the wing held horizontal. This would not produce much lift or forward thrust because the forces on the upstrokes and downstrokes more or less cancel each other out. Sweeping is moving the wings backwards and forwards in the plane of the insect. Again, sweeping with no other movement would be neutral in terms of moving the insect through the air. The third motion is to twist the wing through an angle.

In a typical wingbeat cycle these movements are combined. The wings sweep forward and plunge downwards, tilted down at an angle of about 30–45°, and at the bottom of the stroke they reverse through up to 180° and sweep back again, turning over 180° again at the top (fig. 7.2). The wingtip traces out an approximate figure-of-eight pattern. It is the asymmetry of the downstroke and upstroke that creates the lift and thrust just as a child on a swing moves the body to give propulsion.

The upstroke either achieves some lift (as in the fly) or presents minimum drag (the dragonfly). The wings themselves are not rigid although they do not have any control muscles beyond the root. This means that any flexing of the wing has to be accomplished by a combination of movements by the root muscles and the ability of parts of the wing to 'give' under pressure from the air and thus change shape. Some insects, for example the blowfly, have a subtle additional control mechanism, a miniature gearbox that selects different degrees of leverage for different kinds of flight.

The problem in modelling insect flight is simply stated: in insects the wings move, in planes they stay still. You can put a model plane in a wind tunnel and, using smoke, see the regular patterns of airflow over the wings. The pressures and forces generated can be measured, and the equations of flight calculated. But when a wing flaps, the conditions around the wing change at each stage of the stroke, and the patterns made in the air currents are very complicated.

But, as with the Lotus-Effect and the scanning electron microscope, modern instrumentation, in this case high-speed video, came to the rescue. In 1996, Charlie Ellington, at Cambridge University, capped some 20 years of research into insect flight with a breakthrough study of tethered hawkmoths and a large-scale mechanical flapper that mimicked the hawkmoth's behaviour.

Ellington is a quietly spoken American who came to Cambridge in 1973, intending to continue the studies he had begun on the swimming of fishes, but he was assigned insect flight instead. His office is a museum of biomechanics, because besides Ellington's flapper, now pensioned off, it contains boxes of equipment from an illustrious predecessor, Professor James Gray, who published notable early studies on animal locomotion in the 1930s, especially on the swimming of dolphins.

Ellington published strong papers for two decades which defined the problem of insect flight rather than solved it. In 1984, it was he who showed that standard aeroplane theory could not account for the lift generated by insects, but at the time he was unable to identify the actual cause. Photographs of smoke patterns in wind tunnels looked suggestive but were difficult to interpret because the air patterns generated by the wings are three-dimensional. But all the time the instrumentation was improving.

In 1996, using stereophotography to capture the flows of smoke around his tethered hawkmoths, Ellington saw vortexes – whirls of air like miniature tornadoes – form along the front of the wing on the downstroke to create additional lift (fig. 7.3). To be able to control the experiment better (butterflies and moths are not well behaved in the laboratory) he then built a mechanical flapper about 10 times bigger than the hawkmoth. To preserve aerodynamic similarity, he had the flapper beat its wings at 0.3 beats per second whereas the hawkmoth beats 26 times a second. Models on a larger scale than the actual insects are often used in this work. Michael Dickinson, at Caltech, the leading US expert in insect flight, uses a

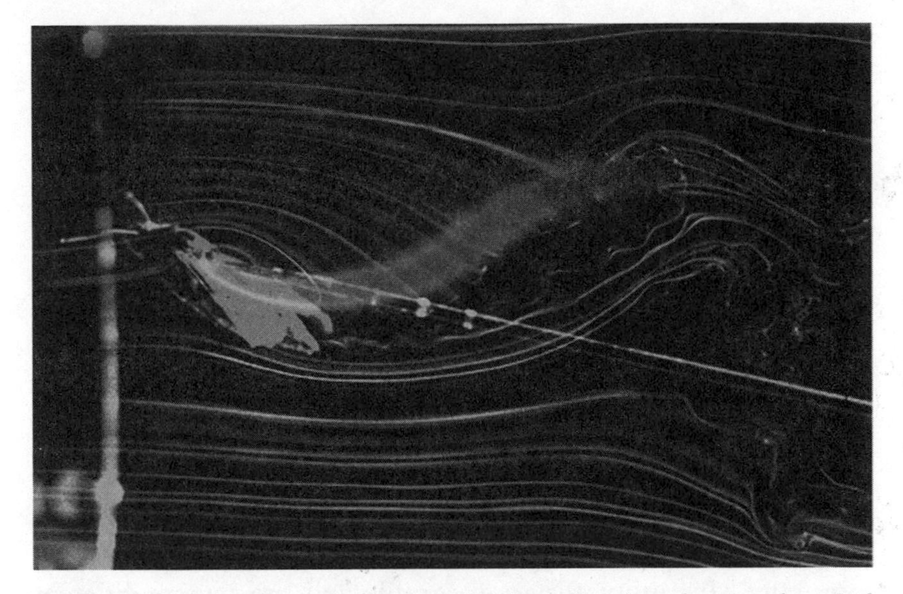

Fig. 7.3 The airflow patterns around a hawkmoth's wing, showing the spiral vortex that produces additional lift, visualized in smoke patterns.

model fruitfly* 100 times life size. To duplicate the fly's aerodynamics on this scale requires the adjustment of several conditions. There is a connection between the size and frequency of flapping of a wing and the viscosity (thickness) of the medium. This means that an insect wing can be modelled by a much larger wing that flaps more slowly in a much more viscous medium, in Michael Dickinson's case, oil.

Although Ellington's work on the flight of real insects was a great inspiration to the MAV movement, he is in the tradition of biomechanic scepticism. One of his first remarks when I met him was: 'We've had so much hype with MAVs – it was ridiculous what happened with them. I lost count but the last time I added up they'd spent $55 million on micro air vehicles and they would have done better to have given $10,000 to a couple of good aero-modellers.' And he groaned. So, although engineers and biologists work together on insect flight, they often see it from very different angles.

* The larger flies all use similar flight mechanisms. Dickinson's studies were done on fruitflies. Fearing's MFI is closer to a blowfly in size but the flight mechanism still applies.

Ron Fearing says of his biologist colleague Michael Dickinson: 'The idea that we could make a flying robot based on his principles is amusing to him and maybe at some point it would provide him with some results that would be important for the things he's looking at, but it's not the thing of real importance, which is understanding the secrets of nature. Michael Dickinson's a great engineer too, when you look at all the apparatus he's had to design just to do his experiments, to get his data, but the engineering is incidental to unlocking the secrets.'

Ellington's work stimulated other laboratories to look at the question and for a while, as so often happens after an initial breakthrough, the position seemed to become more confused. The fact that everyone is studying different insects does not help matters and it is clear that there is no one answer to the question: 'How do insects fly?' In reality, they are versatile and use several different mechanisms. Studies on free-flying red admiral butterflies, a few years after Ellington's work, concluded: 'The micro-air vehicle community may find it daunting that the first flow visualizations of free-flying insects have revealed such a wide range of aerodynamic mechanisms, and that the insect switches between them on successive wingstrokes with such apparent ease.'

But Ron Fearing was certainly not daunted. He has a practical goal and this helps him to cut through some of the confusion. Concerning the various mechanisms he says: 'I think our range of aerodynamic mechanisms is simpler than in butterfly wings. Butterfly wings are so large. You're worried about the compliance [bendiness] of the wings. Our model is the house fly.'

But, given the complexities of flapping flight, is it necessary for an MAV to flap its wings? Fixed-wing planes are not very manoeuvrable and they cannot hover – both highly desirable traits in a small surveillance craft – but when Darpa held a competition in 1999 the winner, the Black Widow from the Californian AeroVironment Inc., was a fixed-wing plane. This led to jokes about Darpa spending millions on feeding the hobbies of overgrown schoolboy aero-modellers (fig 7.4).

What about a micro-helicopter? Both Fearing and Żbikowski reject this and their reasons reveal their biases. Ron Fearing says:

Fig. 7.4 A too zealous copying of nature is not advised. Bio-inspiration requires that the principles be abstracted from nature.

'When you get very small, the ball-bearings in a helicopter rotor are a problem. I wouldn't bet someone a lot of money that they couldn't do it – make a small helicopter – but the fly is the favourite.' For Rafal Żbikowski the problems are that helicopters lose lift whenever they come close to a wall (you don't fly heli-copters into the Grand Canyon); they are also very noisy and have high fuel consumption.

In 1996, Rafal Żbikowski was a young control engineer at Glasgow University when he heard of Darpa's project. He liked the sound of it except for one aspect: Darpa specified both indoor and outdoor use but Żbikowski felt that the gap was in indoor surveillance, both military (think of the cave complexes used by terrorist organizations such as Al Qaeda) and civilian – the D3 jobs (dull, dirty and dangerous). Rescuing people trapped under rubble is the classic mission. Żbikowski says: 'Now they use sniffer dogs, listening devices and thermal cameras. They have to be very careful how they lift the slabs of concrete to avoid a local avalanche. But if there is an opening and someone is breathing, they can fly a small thing in there, have a good look at it and then plan the operation better.'

Żbikowski now works at the Cranfield University campus at the Royal Military College of Science, Shrivenham. Cranfield operates both as an academic institution doing open research and also as a contractor to the UK Ministry of Defence (MoD), giving technological advice and support to the military. Żbikowski is Polish, with a wry sense of humour and a bustling organized manner. Aeronautics on this scale is highly technical and mathematical but when he gives talks he says pronouncing his name is the only hard part.

To him, it is the need for rapid manoeuvrability in confined spaces that tips the balance in favour of flapping flight: 'It's not that I woke up one morning and said, hey, let's do something crazy: flapping like insects. The question is: what is the envelope for indoor flight; how can we realize it with a proven technology, something we know in advance is already working. This is how we reached this conclusion; you just look in your garden and it works.' (Echoes of Feynman there, but this time you don't need a microscope.)

Thinking about insect flight, you really do have to look in the garden from time to time, pinch yourself and say: 'This isn't just a lazy summer's day with bees, hoverflies and butterflies as bit-players around the flowers: this is micro-aerobatics in action.'

Żbikowski's and Fearing's approaches are similar in everything except scale. They are both ambitious programmes and many problems need to be solved. An MAV must be able to sense and avoid any obstacles, take photographs, record data, and find its way back home. To build an MAV you need a tiny power source, a wing-actuating system and a control mechanism.

Ron Fearing began work on the MAV – his version is known as the Micromechanical Flying Insect (MFI) – in 1998 with grants from Darpa and ONR MURI Biomimetic Robotics. His team has taken the approach that since a fly or an MAV has many systems, all of which are essential to achieve autonomous flight, the key systems need to be investigated simultaneously. Since 1998, wing mechanisms, sensor systems and control systems have all made good progress. The wing mechanism is an ingenious contrivance that, if it has not yet matched the fly, comes close in the essentials.

The model is the blowfly, which has a wing span of 25 mm and flaps 150 times a second. So it is quite big as flies go and the high wing speed is the reason that you can hear a fly buzzing but not a butterfly with its more sedate flapping. How do you drive a wing that fast and make it do all those fancy rotations? The best way is to make it resonate.

Resonance is another word for vibration – it simply means something that moves backwards and forwards in a regular rhythm. But it also has a metaphoric meaning – we often say that one thing resonates with another or that a particular experience is 'resonant': this gives us a clue to the deeper meaning of resonance. The idea of resonating with something or someone implies sympathy and on a technical level, as well as emotionally, this means that in order to resonate two things have to be 'in tune'.

You can see how this works with the 'whistling wineglass' experiment. If you run a wetted finger around the rim of a large champagne glass (not a flute), at a certain speed it will begin to whine. It always makes the same note and if you rub faster or slower the note grows louder then dies away. There is just one resonant speed for a particular wineglass and one characteristic note that it emits.

Every structure has a natural resonant frequency – this was the problem with the Millennium Bridge over the Thames – the Wobbly Bridge. Before being modified, if large numbers of people walked in step over the bridge, it settled into resonance at the same frequency as the footsteps – this generated large oscillations, and, if left unchecked, the bridge could have progressively shaken itself to pieces. This fate actually befell the Tacoma Narrows bridge, across Puget Sound, Washington. This was the slimmest most graceful suspension bridge yet built, but its deck resonated in light winds; on 7 November 1940, only four months after it opened, it shook itself to pieces in a wind of only 70 kph.

Not everything resonates. For instance, you can't in any meaningful way make a supertanker resonate like this. But there is another important difference between the supertanker and things that do resonate. When a tanker comes to a stop there is no energy stored to begin the reverse direction: it stops dead. In resonance, the

Fig. 7.5 An impression of Ron Fearing's Micromechanical Flying Insect, 2004.

end of each movement or stroke builds up energy to kickstart the subsequent movement or backstroke, and so it continues.

A yo-yo and a Slinky® are good examples of this. The weight of the yo-yo coming down stretches the string and this energy pulls the yo-yo back when it reaches the bottom (the jerk of hitting the bottom also automatically reverses the rotation). Coming up, the blow when it hits the hand reverses the rotation again and the forces of the blow and gravity send it unravelling down. A Slinky is a spring and it is a combination of gravity and the energy in the spring that keeps it flipping over.

Now it would be possible to devise a mechanism for a fly's wing that operated on the supertanker principle. If energy had to be supplied to bring it to a halt at the end of each stroke and then to start it off again in the reverse direction, a fly would only manage a few strokes before it was exhausted. A real fly operates on the yo-yo principle and so must the artificial fly. Ron Fearing says: 'If it's not running at resonance it means that the flight muscle is doing all this work accelerating and decelerating the wing. The power should be going into accelerating and decelerating the air.'

When I met him at Berkeley in February 2004, Fearing showed me his latest version of the MFI (fig. 7.5). What was fascinating was that there was no attempt to copy the fly's *materials*. Fearing stresses that modern engineering materials such as carbon fibre often have properties in advance of nature's: 'carbon fibre beats chitin,' he says. The wing-flexing mechanism, which was originally steel, is now cut and folded from a sheet of carbon fibre – one of the uses of origami discussed in Chapter 8. The latest version has a seriously thin polyester wing with carbon-fibre reinforcing ribs. Fearing admits that 'the insect wing hinge is very sophisticated – I don't think we've beaten that. But strength to weight the MFI probably beats the fly. The MFI looks big but it weighs less than a fly – the fly's full of water.'

A wing mechanism that can execute the basic figure-of-eight wingbeat cycle is the bedrock but by itself such a mechanism is unstable – the insect has complex control processes to maintain any desired orientation, whether hovering or forward flight or turning, diving or climbing. Planes change direction gradually, partly because they contain at least one person and abrupt turns put a terrible strain on the body: turns are measured in G-force.*

There is a Russian plane – the Su 27 – that amazes airshow audiences the world over with its ability seemingly to stop in the air, and sit on its tail in a blur of shuddering metal and shimmering heat haze, a manoeuvre known as the Pugachev cobra after the Russian test pilot Viktor Pugachev who first performed the manoeuvre in 1989. For a fly, this is routine. Ron Fearing says: 'Michael Dickinson looked at a fly flying in a straight line and making 90° turns. That's about as fast a turn as you can imagine. In less than 10 wingbeats there's a 90° turn. That's one 1/20 of a second to make a complete turn. You'd think there would be something dramatic to account for the right-angled turn, but at first they didn't see anything different between the straight-line flight and the right-angled turn. So they did a very careful sifting of the data and they saw: oh, there's a 1%

* For instance, 2G is twice times the normal force of gravity – a fighter aircraft can pull 9G in a turn.

variation in the wing stroke amplitude. At the moment, building the left and right wings so that they're within 10% of each other is actually pretty good! And now to control it we know we probably need to be within 0.1% if 1% is all it takes to do this dramatic right-angled turn. If you get this wrong, in five wingbeats you can be completely on your back, in an unrecoverable dive. It's a very unstable craft.'

Intriguingly, the 'instability' of insects brings them close to the current technology of high-performance jet fighters such as the Eurofighter and the Lockheed F22 Raptor. In flight, stability and agility are always at war. Insects are designed for agility and planes for stability. It is the need for stability that leads to tailplanes on aircraft: insects never have anything like the tailplanes of aircraft. By definition, stable means resistant to change but fighters need to change their flight pattern in a hurry – that is the essence of being a fighter.

Modern jet fighters can be unstable because, like the insects, they have automatic mechanisms to monitor constantly for instability and to correct it. A pilot would not be able to react quickly enough to do this. So the Eurofighter and the F22 have deliberately unstable flight dynamics, controlled by sophisticated flight-control systems, that enable them to be more agile in aerobatics than previous generations of fighters. It might seem perverse – not to say dangerous – to make a plane deliberately unstable, but it works.

Everyone stresses that insects don't fly by 'deciding' how to move every part of the wing. There are no muscles beyond the roots of the wing so wing motion is dictated by the force and direction of these muscles, the bendiness of different parts of the wing and the force of the air acting on it. The fabric of the wings is so made that they adjust their shape in ways favourable to the motion under way. A simple example: if a sheet of material is curved like an aircraft's aerofoil, it changes the way it flexes. You can model this in paper: take a sheet of A4, bend it slightly in the lengthwise direction as if you're going to roll it up into a cylinder. If you grip one end and flap it, the sheet stretches taut on the downstroke and bends on the up. This is far too simple as a working wing but it does behave very differently on the up and downstrokes, which is the first thing that you need, pushing

Fig. 7.6 The Shrivenham insect-like flapping mechanism.

hard against the air on the down, taking the line of least resistance on the up. If wings did the same things in both directions then the forces would cancel out and there would be no lift.

One big question for MAV developers is: how much of the insect wing do they have to copy? The hinge where the wing joins the body is relatively long in most insects, restricting the amount of bending; by making the wing bendy enough to twist in the upstroke, despite the rigidity of the base, the wing can achieve the necessary profile on the upstroke. In any case, if the fly is the model, with its very small wing, too much twisting is not desirable. As Ron Fearing says, if the wing were too bendy 'it would be like paddling a canoe with a cardboard paddle'.

At the 15 cm scale Rafal Żbikowski is working at, a flexible wing is necessary and his team have shown that quite simple synthetic wings, with only two control 'muscles' at the root, can bend in ways convincingly like the real thing. Żbikowski has a test bed model of his wing system, in which an ingenious linkage mechanism generates the figure-of-eight pattern of flapping flight (fig. 7.6).

For both Fearing and Żbikowski, the wing-flexing system,

Fig. 7.7 The head of a fruit fly under the microscope looks like a highly tooled piece of engineering. The eyes give almost 360° vision.

although complex, seems a manageable problem. The big question is control. The MAV is a self-steering *robot*, which means that it has to sense its environment and give appropriate control signals to the wing flexors. The more we know about insects' ability in this department the more impressive they seem.

The fly's sensory system is highly developed. Under the microscope, the head of a fly looks like a highly tooled piece of engineering (fig. 7.7). In their large compound eyes, the smaller species have several hundred and the larger ones several thousand individual eyes, each one of which views a tiny section of the visual field. The eyes give almost 360° vision and there are additional aids. There are three light-sensitive cells called ocelli on the top of the head which give the fly a constant sense of up and down, so that it does not become lost in a tumbling 360° maze. Antennae and wind-sensitive hairs also abound.

The fly's neatest control mechanism is probably the halteres.

These are a development in the larger, more advanced flies (the ones we are most familiar with) of the hind wings, which have become adapted as a gyroscope.* Staying upright is a major problem for an unstable flying platform with wings beating 150 times a second. The halteres beat out of phase with the wings and resemble pendulums, giving the fly a constant reference point for its motion in three dimensions. In the engineered 'fly', the halteres can be mimicked by miniature gyroscopes, with photocells serving as ocelli.

Each one of the fly's hundreds of compound eye elements experiences a directional flow of sensations during motion – the scientists call this optic flow. The nervous system and brain need to analyse this data to deduce the fly's position and direction of motion relative to its surroundings. Flies can process a prodigious amount of data: 17–18 pictures a second give us the appearance of continuous motion; the fly needs 150–200 pictures per second, the same speed as its wingbeat. This is one reason why flies are so hard to catch: they perceive and act upon the world far quicker than we do.

The basic theory of this kind of motion detection was worked out as long ago as the early 1950s. And there are computer simulations of how the fly uses the optic flow to correct deviations from a straight-line course. This is one important aspect of a fly's repertoire – others, such as landing, are more complicated.

Rafal Żbikowski explains that a major difference between insects and aeroplanes is that aeroplanes have few sensors whereas insects have hundreds of them. The central control for an insect's flight is relatively crude (this is where the less computing power than a toaster comes in), so it relies on local feedback loops between these many sensors and the actuators. A modern aircraft such as the Eurofighter relies on only some 20 different measurements feeding in to its control system, compared to 80,000 for the fly. If you have 20 measurements you can only have 20 different independent feedback loops, and the plane's on-board computer has to solve some very complex equations. But the insect has so many feedback loops it does

* Many insects have four wings, the dragonfly being the classic case, but the hind wing is modified in several insects, although not to the same degree as the halteres of the fly.

not need to solve these equations. Żbikowski stresses that *if you can measure everything you don't need to compute so much*, and this seems to be the way that insects do it. It is a little bit like the way we execute a skill such as throwing a ball. We don't have to solve equations to do it, the way that digital computers do: we look at where we want to send the ball and the feedback systems between our nerves and muscles, based on past experience, do the rest. Żbikowski believes that this sensor-rich feedback control system will be the key to controlling MAVs.

So how long will it take for all the components to come together and an MAV take to the air to execute a few manoeuvres under its own power?* Rafal Żbikowski says: 'In five years we will have something that is mechánically viable as a flapping device, ie, a reference platform that will have a chance of getting airborne.' Ron Fearing's mechanisms have operated statically in wind tunnels, and one has produced motion when attached to an arm like a record player stylus. As Fearing says: 'You could make an MAV yourself. You can buy a rubber band powered flapper for $10.' But a rubber band only lasts seconds: they are aiming for the real thing.

Ron Fearing believes that this is not too far away: 'We're starting to find that robots can be smarter than insects – in the eighties it was clear that the insects were more capable. I hope in this decade we'll see robots being more capable than insects.' The current target for an MFI flight is five minutes, limited by the battery. But Fearing points out: '5 minutes = 300 seconds, at 3 m/s gives a kilometre range. In a

* Darpa has a history of being over-optimistic about such projects: in March 2004 they organized a challenge competition for wheeled robots to cover a 240 km course in the Mojave Desert. The incentive was $1 million but carrots like that only work with humans, not robots. Of 106 putative entries only 13 were ready to start; half got lost in the first mile and the winner, a converted humvee from Carnegie Mellon, covered a mere 11 km before its front wheels caught fire! Darpa were not fazed: 'We're going to run it again in a year and the prize money will be $2 million,' said a spokesperson. Before the race, Ron Fearing said: 'None is going to make it this year. Maybe in five years. But the question to look at is: given these big gasoline-powered robots, how long can a robot go independently before it gets stuck and needs some help? Then think of having a fly robot that can go for an hour ... the world's a pretty complicated place.'

building, or an urban environment, you can't go that far without being out of radio range. If a building is on fire, and you can't find victims in 5 minutes, the fly would not be that useful. Having it last one hour in this scenario would not greatly increase its utility. Alternating between perching 95% of the time and flying 5% is a good strategy for longer duration missions. Perching doesn't take much battery power.'

The MFI is the closest thing on the stocks to those swarms of nanobots Michael Crichton let loose on our imaginations in *Prey*. Indeed, Fearing does talk about releasing them in swarms, the point being that each individual MFI is expendable and it wouldn't matter if it did not complete its mission. Only one needs to get through. But if an MFI gets out of control it is not going to unleash the mayhem seen in *Prey*: it will merely flop onto the ground, spent. The MFI is a small machine, made from carbon fibre and polyester, some steel and other bits and pieces. There is no way such a creature could be made from biological material in even the distant future. But now we can see what it really does take to make a small, self-controlling flying vehicle, we have to admire the originals – the flies, moths, butterflies and bees – all the more for it.

CHAPTER EIGHT

Origami for Engineers

> 'At a slight angle to the universe.'
>
> EM FORSTER on poet CP CAVAFY

Nature likes things that can fold away into something small and open out when they are needed. Engineers call these things deployable structures, which sounds very grand but we all know about deployable structures: the umbrella is the perfect example. When a button is pressed, the shaft expands and in doing so forces open the array of spokes that carry the fabric. But an umbrella illustrates the key problem with deployable structures: it is not fully automatic – when it opens it has to be locked into place. And once the spars open, the fabric starts to tighten and the force required to lock the mechanism increases dramatically. Also, it has a kink in its operation: once locked, it requires a major jolt to unlock it. It is what engineers call a bistable system: flipping from one state to another is quite difficult. In other words, it is hard to design a system that automatically goes up and comes down again.

A beetle needs to fold its wings away under the wing covers when on the ground, but to deploy them in a hurry for flight if danger threatens. Over the course of evolution, nature has had to come up with deployable structures that work reliably every time. And she had to make them as easy to take down as to put up. There is no one to give that beetle a jolt if its wings happen to get stuck. Other than a predator, of course.

Fig. 8.1 The Ha-ori leaf-folding mechanism proposed by Biruta Kresling, and a real hornbeam leaf.

So nature is pretty good at folding and, until recently, human folding techniques were exceedingly unimaginative by comparison. If we consider our normal folding techniques – that is everyday folding, not advanced origami – we fold things at right angles, and whether it is newspapers, maps or sheets, if a fold is not straight the whole thing goes wrong. But there is a problem: maps might look tidy folded up but, once open, if you get the folds out of sequence, they won't go back again. Imagine instead a map that you could open and close with a single pull.

In *As You Like It*, Shakespeare eulogized nature for containing 'tongues in trees, books in the running brooks, sermons in stones'; he might have added 'magical map-folds in the leaves' because the one-pull map *can* be made and it works in the same way that some leaves, especially those of hornbeam and beech, unfurl from the bud. A leaf bud is much smaller in both length and breadth than the leaf it will become (fig. 8.1). After bud break the leaf will begin to grow but even at birth the leaf is too big for its cocoon. How can something be so tightly packed and then expand in both length and breadth as it opens?

The principle of leaf opening can be demonstrated in 10 minutes with an origami experiment known as Ha-ori (meaning 'leaf-fold') (fig. 8.2). The 'leaf' expands in both length and breadth when it opens, and although it won't quite close simply by pushing on the ends, it makes a good attempt. (The map-fold, as we shall see, takes this all the way.) The leaf-fold also gives a good demonstration of 'bistability'. If you pull very hard on the ends, the pleats flatten out and the whole leaf flips into a reverse curve. It is now stuck like an opened umbrella and is hard to refold until the kink in the spine is removed and the pleated structure recreated.

This leaf-folding pattern was discovered by the German-born Paris-based designer Biruta Kresling in 1990–1, after meeting Koryo Miura, the Japanese expert in folding structures, at a symposium in Budapest in 1989. From him she learned the paper-fold known as Miura-ori (*see* page 187). When she returned to her job of teaching design at Valenciennes, she was looking for ways of getting her students to explore new ways of folding. There were hornbeam trees on her way to the college and she was struck by their strongly pleated appearance. Could they be modelled with Miura-like folds? They could, and the technique showed how they could be tightly folded in the bud and emerge both longer and broader than their cocoon.

Kresling presented the leaf-fold at symposia and art exhibitions but the idea entered mainstream science in 1997 when she was at Reading University, the home of English bio-inspiration, working with Julian Vincent. Together with Hidetoshi Kobayashi, who had studied at Reading with James Gordon, they developed the idea,

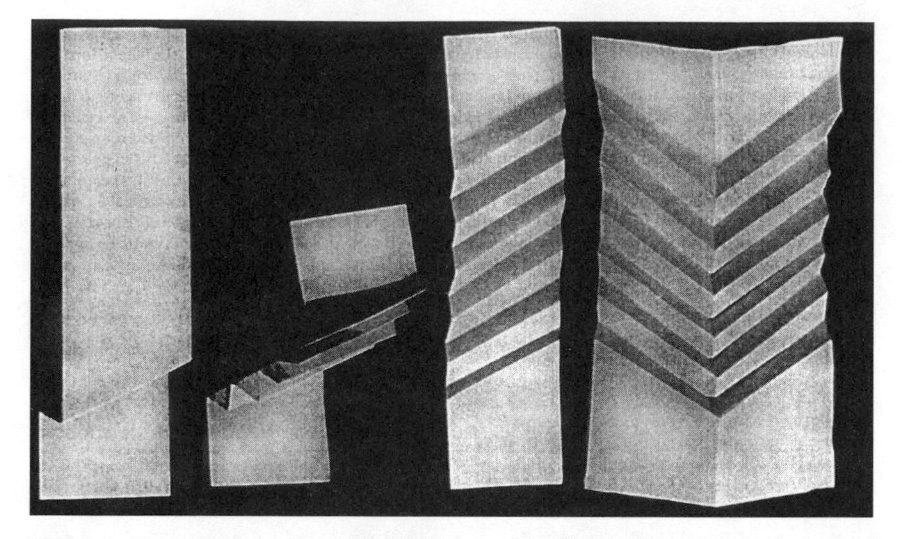

Fig. 8.2 How to make the leaf-folding mechanism. Take a sheet of A series paper (it is best to start with A5 because although A4 makes a very impressive leaf, it is more demanding) and fold it in the middle lengthwise. Take the folded sheet and about 3–4 cm from the end make a fold at about 60° from the long axis. Turn the sheet over and 1 cm further on fold back parallel to the first fold. Then turn it over, fold another 1 cm on, making sure to maintain the parallel structure. When you reach the end, open the paper out. The ribs now look a little like a leaf, but it won't yet fold up because the valleys and ridges on the two sides are out of phase with each other. So all the valleys on one side should be refolded as ridges and the ridges as valleys, leaving the other side as it is. The ridges and valleys now make pleats and the whole is a flying-V shape. By pinching from the end, this can be coaxed into folding up into a bulky, flat folded strip. Now pull the ends and the leaf opens.

---

worked on the mathematics of it, and published the results in the *Proceedings of the Royal Society*. The leaf-fold is an excellent educational tool and it was used in workshops at the Science Museum in 2000, under the banner 'Just Fold It!'

Nature invented Ha-ori but we became aware of it through the medium of origami and, in particular, through the work of Koryo Miura. Miura is a Japanese space scientist and exponent of serious origami. He was, until his retirement, Director of the Institute of Space and Astronautical Science in Tokyo and his revolutionary work in origami has been influential in the very different fields of biology, cartography and space science. Miura's ideas on folding took shape at NASA's Langley Research Center, Virginia, USA, during 1966–7 and initially developed from an abstract question: What happens if you fold a thin sheet of paper in two directions at once?

The crude way of shrinking a piece of paper in two directions at once is to grab it in your fist and scrunch it up. It is now a ball in three dimensions rather than a sheet of paper in two! While this may not appear very helpful, an ordered form of crumpling is exactly what Miura was thinking about. Such ordered forms of crumpling were known from experiments in crushing metal cans with pressure from the top. Instead of the mess you get with scrunched-up paper, the buckling pattern shows strong regularities: diamond-shaped patterns appear and the material buckles both in and out of the plane surface of the cylinder. This is known as the Yoshimura Pattern.

If a can with the diamond crush pattern is rolled out flat, a sheet with suggested regular fold lines becomes apparent. The lines mark out the best crumpling pattern for the sheet. This evokes George de Mestral in his imagination rolling out the bur structure to create the hook-and-loop fastener. The secret is to build these diamond-fold patterns into the paper to make it easy to fold up.

By 1978, Miura had a general theory of how to collapse a sheet. He came to two conclusions: 1) that it could be the answer to his problem of erecting solar-panel arrays in space, and 2) that it would make a good map. He decided that the map was a quicker way to demonstrate the concept than the space array.

As Miura says: 'Although map folding is simply good fun, it provides much useful information, which pure mathematics cannot do.' To make a map, the diamond patterns become almost rectangular, and this is the key to the easy foldability: the folds work best at about six degrees off a right angle. The map instantly revealed that because all the folds are interdependent, the entire map opens – and even more miraculously closes – with a single pull from the front and back corners.

Miura initially described his technique as a 'Developable Double Corrugation Surface' but the process was rescued from the eternal obscurity this name would have brought it by the British Origami Society. The map idea was launched at the 10th Conference of

International Cartographers in 1980 and it created a buzz, despite its forbiddingly technical name. The British Origami Society gave it the more poetic name Miura-ori (Miura's Fold), which pleased Miura and everyone else.

For this act of charity Rupert Bear is partly responsible. In England, many people's first encounter with origami (although it wasn't called that then) was through the *Rupert Bear* Annuals. Rupert Bear's creator Alfred Bestall was an early exponent of the art of origami, which had been invented in Japan in the 1930s (there had been paper folding in Japan for centuries but everyone agrees that there was a new beginning in the 1930s). Bestall's Rupert Bear origami patterns began to appear from 1946 and continued up to 1985. With Robert Harbin, the magician, Bestall founded the British Origami Society in 1967: in good time to bestow upon Miura's fold its more evocative name.

To make a Miura-ori one-pull map involves creating a pattern of repeated folds similar to the leaf model (fig. 8.3). Most people are sceptical of the idea until shown that it really does work. Miura-ori is a piece of practical Zen (the rustle of one-pull opening). Like Velcro and Catseyes, it is a low-maintenance way of doing something that would be clumsy by other means. Miura-ori is the answer to the problem of how to make closing as easy as opening.

Unlike the umbrella, Miura-ori isn't bistable: the open and closed positions are on a continuum with the corrugated surface you see when it is half open. And when it is open, it is not really flat: the folds are just at very shallow angles. So there is no awkward kink in the mechanism to jam it – if it were perfectly flattened once opened then it would be harder to close.

Miura-ori is obviously a better way of folding maps so why aren't we all using them? Well, a lot of people are: or rather they are using the closely related Z® Cards, which have been phenomenally successful all over the world. The Z Card is the brainchild of George McDonald, a resourceful inventor-businessman, who as a prolific travel writer in the early 1980s became frustrated with old-fashioned maps that once open were difficult to refold neatly. In many parts of the world, tourist boards would put out maps that were single A4 unfolded sheets and McDonald was forever folding them up to

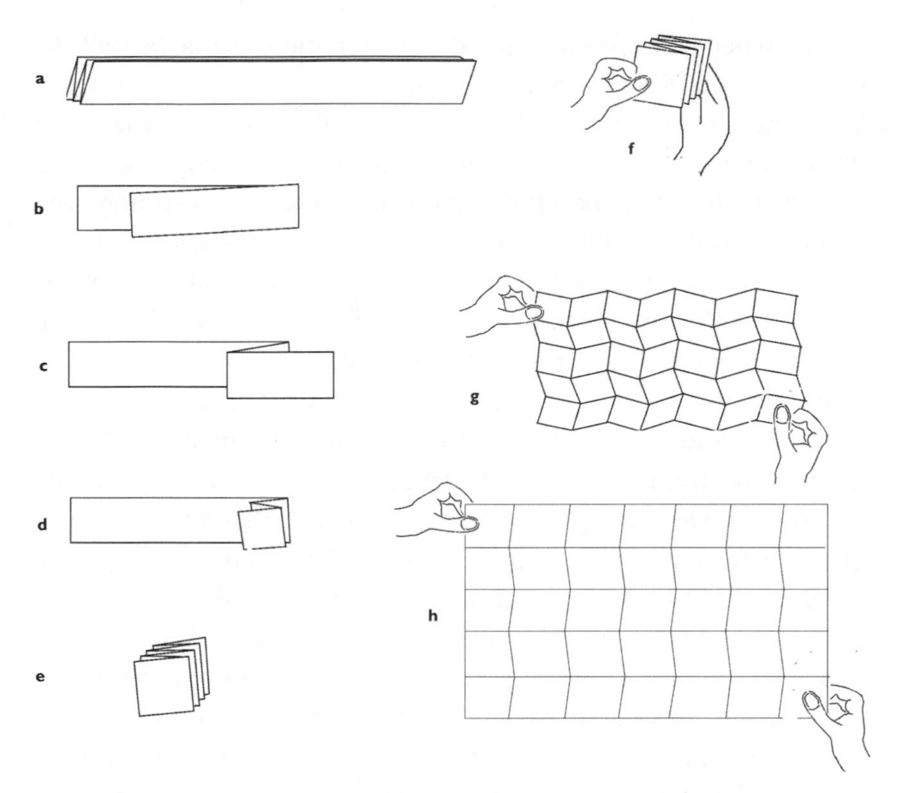

Fig. 8.3 How to make a Miura-ori map. a) Take a sheet of A4 and mark it out in five equal strips lengthwise. Fold along the lines accordion-style until it is a single narrow strip. b) Fold the end of this strip in the ratio 4:3 but skewed at an angle of about 6 degrees from a right angle. c) Fold the little end back and parallel to the original strip to make a skewed square (the sides will be approximately 42 mm). Fold up the short end into three equal cells (d) and then do the same with the long end – four cells. You will now have a single map with the cells slightly skewed so that they overlap (e). There may be a little uncertainty at first as to how long the folds should be because the lack of right angles is initially a little unsettling. But after a few attempts you will get the feel of it. f) When the concertina is complete, grasp the end sheets only and pull. The whole sheet opens out in one go (g and h). As a finishing touch, cards glued to the ends make a map more durable and even easier to open. Once you have got the hang of the small A4 map you can tackle larger sizes.

---

credit-card size to stow them in his wallet. He realized that there had to be a better way of doing this and researched the map patents extensively before coming up with the Z Card fold. It is so called because when half open it forms a characteristic Z shape. In 1985, he took out a patent on the one-pull map.

It is the direction of the folds that is critical for practical maps: the off-90° angles of Miura-ori *facilitate* opening but paper can flex enough to give the bending needed to keep the folds working even with conventional right-angled folds. With larger maps, Miura-ori has an advantage because the angled folds more naturally assume the intermediate stage. In the first stage of opening, a Miura map typically falls into five leaves while the Z Card has three.

Miura's map is easy to make as a prototype by hand, but to get a machine to fold these angles and to assemble the map automatically, rapidly and cheaply, is a major challenge. The Orupa company solved the problem well enough to bring the maps to market in Japan in 1999 but they are still expensive compared to conventional maps. The perfect Miura-ori mapmaking machine still awaits its inventor.

The Z Card is cheaper to make than the awkwardly angled Miura map, but it still took seven years to bring the Z Card to market, and mechanizing the folding was quite a problem. Z Cards are not just maps: they are used for portable information of all kinds, being more user-friendly than fliers and leaflets. Many Z Cards are credit-card sized and the maps have followed the credit-card explosion around the world, always selling best in countries with high credit-card usage. When I was in San Francisco, researching this book, the local map provided by my hotel was a Z Card. Like the lotus cake in Seattle, this seemed a good omen.

The story of Miura-ori and Z Cards is a classic tale of two ways of inventing. Miura, the scientist and technologist, was looking for ways of folding space antennae; he started with an abstract principle and found something so universal that it connects tree leaves, maps, crushed cans and space arrays. George McDonald had a practical map in view from the start. He acknowledges Miura as the master map-folder and thinks his maps are works of art. In Japan, Z Cards and Orupa collaborate on maps. Z Cards still manufacture in Britain, at Warrington, and George McDonald talks of the successful inventor's burden: instead of thinking how best to fold a map he loses sleep wondering if he has gained enough orders to keep the factory working at full capacity.

Miura-ori is a whole family of related foldable structures and there is probably no end to the variations on this theme. Simon Guest, who works on deployable structures at Cambridge University, has an interesting variation in which several 'leaves' radiate to create a circular folding mechanism. There are two patterns: leaf in and leaf out.

Leaf out makes a handy one-pull map for maps smaller than the classic Miura-ori. It is good for city centre and underground railway maps. Something very like a simplified version of the leaf-out pattern is used by Compass Maps Ltd in their Popout maps. Unlike Guest's design, the Popout is rectangular rather than square but although it lacks the corner folds, it folds up in the same way.

For Miura, these maps were a by-product of his work throughout the 1980s and 90s developing solar panel arrays for space satellites. Solar panels need as much surface area as possible to maximize the energy they produce from the sun's rays. But to be launched by rocket they need to pack down into the smallest size possible (in this they are very similar to leaves in that they also aim for maximum exposure to the sun and pack down tight in their launch module: the bud). Solar panels are usually extended in only one direction, but these arrays, like the leaves and the map, expand in two dimensions, greatly increasing the surface area. In Miura's space arrays, the panels have the same shape and angles as the paper-folds but since there is no one in space to pull the ends they need joints and tension struts to manipulate them. Such an array went into space for the first time on the Japanese Space Flyer Unit on 18 March 1995, and was retrieved by the NASA Space Shuttle *Endeavour* on 13 January 1996 (fig. 8.4).

Miura's Space Flyer Unit was a one-off to prove the concept: one-dimensional arrays are sufficient for the relatively small solar panels currently required but Miura-ori is being developed for solar-sail propulsion units for spacecraft. The solar sail is a visionary idea for long-distance space travel, beyond the solar system. The idea is to use the solar wind, the rain of particles of light emitted by the sun, to accelerate the spacecraft. The pressure is very small, but with a large sail it should be enough to produce acceleration. In the near vacuum of space there will be no brake on this acceleration, so very high

Fig. 8.4 Two applications of Miura's folding patterns: a) Japanese drinks cans made by the Toyo-Seikan Company use the pre-folded pattern known as Pseudo-Cylindrical Concave Polyhedral (PCCP), a close relative of the map-folding techniques; b) Koryo Miura's Space Flyer Unit, powered by a solar-panel array which folds up using the origami technique of Miura-ori.

speeds should eventually be attained. The first solar-sail spacecraft is due to be launched in 2005 by the Planetary Society, a private body, using a Russian submarine launcher. This first sail is not a Miura-ori but a fairly simple eight-panel, 30 m diameter structure. Size matters with solar sails and the use of Miura-ori will allow larger sails to be deployed.

Another practical application of folding is Julian Vincent's bio-inspired camouflage, referred to in Chapter 5. This smart camouflage, based on the colour-changing properties of the squid and cuttlefish, is designed to be deployed from vehicles using a mechanism inspired by the leaf. This raises the prospect that troops might one day go into battle under the cover of squid camouflage erected by a beech-leaf mechanism.

To return to those crushed cans that inspired Koryo Miura, they proved to have a practical application slightly easier to implement than the map. Miura realized that if you could create a cylinder with these diamond patterns already built in, part in and part out of the plane of the cylinder, it would be stronger, more resistant to buckling, than a standard can. The pattern is called Pseudo-Cylindrical Concave Polyhedral (PCCP) – this time there was no British Origami Society to give it a better name. The pattern shows strong affinities with Buckminster Fuller's geodesic domes (*see* Chapter 9) and the bacteriophage (*see* Chapter 6).

In the 1990s, the Toyo-Seikan Company introduced both pressurized and vacuum-packed PCCP cans (fig. 8.4). Pressurized cans are used for beer and fizzy drinks: the pressure blows out the polyhedra so that when unopened the can appears to be a cylinder with faint diamond-shaped fold marks. The moment the can is opened, it collapses into the pre-folded PCCP buckling pattern like a controlled Yoshimura collapse. For vacuum-packed products, especially tea drinks, the PCCP pattern is fully developed: the can has prominent diamond-shaped facets. The pressure now comes from the outside and the can resists collapse by being partially 'pre-collapsed'. In both cases, the strength of the cans allows 30% less metal to be used than in a standard can.

The link between the leaf-folds, the cans, the maps and the space arrays is that all these surfaces are doubly-corrugated: while being basically flat or cylindrical, they have built-in angular folds that enable folding (or, in the case of the can, resist folding) in a plane at right angles to the original surface. The leaf-fold shows us that it is not only on the nanoscale that nature uses surprisingly subtle methods beyond our old-fashioned right-angled engineering.

It is not only Miura's maps that have a Zen quality: cans that don't collapse because they are already pre-collapsed and gossamer solar sails that fly across the universe by the pressure of light alone have an equal charm. Such things *could* have been invented anywhere but it seems appropriate that they should be Japanese. Biruta Kresling regards Koryo Miura as her master 'in the best Zen sense'.

Miura-ori is not the first example of paper folding coming to the aid to engineers. In his classic *Structures: or Why Things Don't Fall Down*, James Gordon, the founder of bio-inspiration in Britain, tells the story of how honeycomb panels came to be used in aircraft construction. In 1943, a circus proprietor called George May came to see him with a new turn. It was a paper honeycomb structure, rather like a party paper-chain decoration, and May wondered if it might be of use in aircraft construction. As it happened, Gordon and his colleagues were looking for something just like this: Gordon realized that stiffened with phenolic resin, the honeycomb would be strong enough for aircraft construction and very light. It was widely used as such during the war, and now finds its main use in the standard cheap panel-doors used in houses and offices. Gordon was well aware that if you want a light structure, it is a good idea to fill it with holes. He might have gone straight to nature for the honeycomb idea but it was the paper chain that sparked it off.

Origami is taken increasingly seriously by engineers. It is recognized on the syllabus at some universities in France, and the International Meetings on Origami Science, Mathematics and Education are a forum in which scientists and mathematicians can debate the serious side of paper folding. It is also possible to learn about the wing-folding mechanisms of insects through origami. The wings of insects need subtle folding techniques for two reasons: in many insects the wings need to be folded away when not flying. And they also need to use folding techniques to cope with the demands of flight. As explored in the last chapter, insect wings change shape during the wingbeat cycle but there are no muscles within the wing, only at the root. This means that they need to use air pressure to shape them in the desired ways. If they were floppy in every direction they would always bend away from the air current, so losing force. They resist this by having structures that bend the wing in response to air pressure. This is rather like the principle of the PCCP cans: resistance to folding can be achieved by building in pre-folds along the lines of folding stress.

Insects have some ingenious mechanisms to *increase* the curvature of the rear edge of the wing during the downstroke and they can be modelled in paper or card. Firstly, if a piece of paper is curved lengthwise it is harder to bend it upwards than it is to bend it downwards. If the curve is arranged to be off-centre, with a sharper curve at the front, as insect wings are, the rear edge tends to bend down while the front edge stays straight. The front edge of insect wings is stiffened, with thicker ribs than the rear.

But although such a rear edge can bend downward easily, there are no muscles to do it; while the wing is beating down the air is pushing up, so how can the rear edge be forced *down* without any external force? In one model, the umbrella model, when the wing pulls down, the outer margin is pulled tight like a drawstring, thus compressing

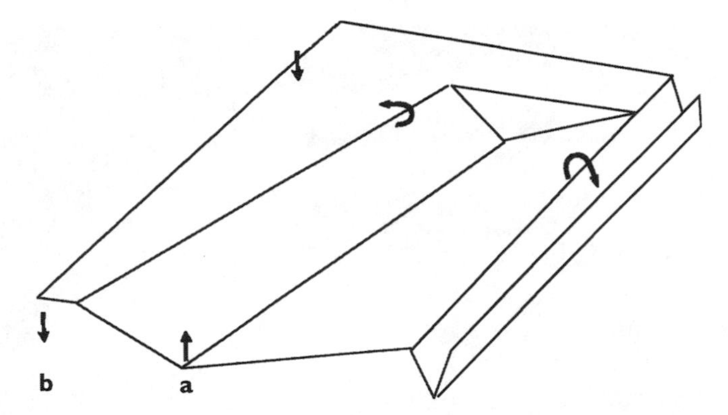

Fig. 8.5 Insect wings bend to enhance their aerodynamic performance thanks to a kinked structure called the 'arculus'. This can be modelled in card. Upward pressure of air (a) as the wing moves down depresses the trailing edge (b) to enhance performance. The leading edge is stiffened and so does not move in the same way.

---

the rear wing veins and causing the wing to bend. The locust uses this model.

Dragonflies use another technique: the wing is not flat but it has an out-of-plane triangular kink. This kink makes the wing behave asymmetrically in the downstroke, depressing the wing's rear edge (fig. 8.5). The wings of flies are less bumpy than those of dragonflies but they also have a triangular kink. The paradoxical movement of these wings, with the rear edge moving down during the downstroke, is related to Miura-ori. It is because the structural kink in the wing is out of the plane that the rear edge depresses when the air pressure is trying to force it upwards.

Origami is not just useful in modelling insect wings: it also plays a role in micro air vehicles (MAVs) too. The wing-flapping mechanism of Ron Fearing's Mechanical Flying Insect (*see* Chapter 7) uses origami. The carbon-fibre structure is cut from a single sheet and then folded.

Origami is all about folding in two and three dimensions but nature has some ingenious deployable mechanisms that extend in one dimension only. One of the most dramatic examples is the single-celled organism *Vorticella*, first seen by Leeuwenhoek in 1676. *Vorticella* is a protozoan, which means that technically it is not

Fig. 8.6 The Millbank Millennium Pier, serving Tate Britain in London and designed by Marks Barfield Architects, has an origami-based structure that uses the minimum of materials for the job in hand.

regarded by biologists as either an animal or a plant: it belongs to a third way of life on Earth. However, it does feed on bits and pieces rather like an animal instead of using chlorophyll to make its food, so it is more animal-like than plant-like. It is normally attached to a leaf by a stalk 2–3 mm long but it can contract on an instant so that the squat cellular body of the *Vorticella* sits on the leaf and the stalk seems almost to have disappeared. In fact, it has become a very condensed rubbery mass.

It can then re-erect the stalk, although more slowly, in a few seconds. When the stalk is extended the filaments are held apart by electrical repulsion. The collapse is caused by positively charged calcium ions which neutralize the electrical repulsive forces. To re-erect the stalk the creature is able to quarantine the calcium ions, bundling them up inside the cell where they can do no harm. If it were possible to scale up the mechanism of this miniature ionic umbrella and make it work with the ions in rainwater, the ionic umbrella would remain furled in dry weather and with the first drops of rain would instantly deploy.

Besides its use in bio-inspired mechanisms, origami has a more metaphoric use in larger structures. Origami-like forms can be created to satisfy design and aesthetic criteria without actually being made from folds in a single material. Computer-aided design and fabrication makes it possible to design very complex shapes that would have been expensive and difficult to fabricate with traditional technology. This results of this process are seen in the Millbank Millennium Pier (fig. 8.6), designed by Marks Barfield Architects and opened in May 2003 to serve Tate Britain and connect the two Tate galleries. The shape chosen by Marks and Barfield is an extraordinary asymmetric collection of wedges, inspired by origami and revealing different facets from every direction.

Steve Chilton, Marks Barfield's designer for the project, says that despite the pier's playful appearance, 'There's not a single fold I can't give you a reason for.' What this means is that the pier is faceted, not just for its inherent visual appeal, but for the same reason that the head of a bacteriophage is faceted: it is the minimum-surface option. The pier is essentially a route for getting people on and off the boats. Once their standing room has been satisfied, material can be lost from parts of the structure that don't impinge on the envelope carved out by walking passengers. Faceting is the answer. There is hardly a right angle in the whole structure and, in one corner of the waiting area, the end wall seems to continue in a plane with the side wall, despite actually having turned the corner.

The Tate Pier has obvious visual affinities with other structures in this book. It is, in some aspects, very like a phage, a Stealth bomber or Ron Fearing's MFI. It is also a piece of architecture, the largest kind of structure. And this brings us to our final topic: large-scale architecture employing organic principles. We have come a long way from nanostructures, but nature is still the touchstone.

CHAPTER NINE

The Push and Pull Building System

> Steady under strain and strong through tension,
> Its feet on both sides but in neither camp,
> It stands its ground, a span of pure attention,
> A holding action, the arches and the ramp
> Steady under strain and strong through tension.
>
> SEAMUS HEANEY, 'The Bridge'

Although the bulk of bio-inspiration comes from tiny forms, we were builders long before we became nanotechnologists and alongside the growth of bio-inspiration in materials science we have seen the rise of a new organicism in architecture. This achieved public recognition with the *Zoomorphic* exhibition at the Victoria and Albert Museum in September 2003.

Zoomorphic showed that the roll-call of architects using bio-inspiration in at least some of their buildings is impressive, including Will Alsop, Edward Cullinan, Norman Foster, Frank Gehry, Nicholas Grimshaw, Marks Barfield, Renzo Piano, Moshe Safdie, and, the most consistently organic of them all, Santiago Calatrava. The roll-call of animals is equally impressive and wide-ranging: armadillos, sea sponges, shrimps, rays, vultures, dragonflies, trilobites, beetles, jellyfish, and so on.

Admittedly, some of the more fanciful buildings, such as Birds Portchmouth Russum's Morecambe Bay development, with giant shrimps standing sentinel over the wide bay, never got beyond the paper and model stage. Nevertheless, there are enough buildings and projected buildings worldwide to make it likely that bio-inspired architecture will come to be seen as *the* style of the first decades of the 21st century.

If the relationship between bio-inspiration and technology is not always clear-cut, the larger the structures the greater the uncertainty. The *Zoomorphic* exhibition stimulated discussion between architects and scientists on how deep are the roots of the organic tendency. In September 2003, at the annual Biomimetics Conference at Reading University, Julian Vincent, Professor of Biomimetics at Bath University, was scathing about what he sees as an often superficial appropriation of the outward forms of living structures without learning from the way that nature actually functions. At its worst, he says, the architect's approach comes down to: 'I'll say I got the structure from an animal. Everyone will buy one because of the romance of it.'

What Vincent would like to see is architects learning the *principles* of natural structures. He gave the example of the air-conditioning system perfected by some species of African termites. The termites need to keep their temperature and humidity within narrow limits and to do this the mounds, which are built from faeces and spoil (on the ancient Chinese dungheap principle; *see* Chapter 6, page 135) and can be 10 m high, have a ventilation system which uses the wind and narrow passages running up the mound just under its surface to maintain the right conditions.

The termite principle has long been used in vernacular architecture in hot countries such as Iraq where summer temperatures of 50°C are common. In the traditional Iraqi house, roof-top vents called *badgirs* draw cool air up from deep cellars through air passages running up the outside wall. The idea entered contemporary architectural practice with Pearce McComish's Eastgate Building in Harare, Zimbabwe (1996). Eastgate consists of two buildings astride a central atrium. The windows have large horizontal concrete shades and the atrium stays cool, sandwiched by the two flanking buildings. Cool air from the atrium is circulated up the voids in the outer walls of the offices. Eastgate is a success but Vincent believes that the vernacular technique is still superior to the designer version.

Architects are now concerned to design buildings that use energy efficiently: the spiralling latticework of Norman Foster's Swiss Re Tower (2004) resembles the cage of the Venus flower basket (*see* page 128), but its organicism is more than skin deep. The internal spiral wells provide natural ventilation and the projectile shape reduces the howling gales that roar around the base of conventional slab skyscrapers.

But most of the buildings featured in *Zoomorphic* use nature in a less functional way, and Julian Vincent's conclusion, that 'architecture has a hell of a long way to go before it uses what's available in biomimetics', is obviously true. How could it be otherwise in such a new discipline?

The connection between small-scale work such as trying to model the architecture of the gecko's foot and modern building techniques is *form finding*. When architects and structural engineers resolve the forces in the structures they want to make, these are the same forces that nature had to contend with as her own structures evolved. And if nanotechnology is all about size and shape so, on a different scale, is architecture. This is why a photograph of a structure such as the *Lychnis* seed coat (*see* page 33) looks like advanced human architecture.

In *On Growth and Form*, D'Arcy Thompson wrote about early links between building and nature. Giant trees have to withstand large forces because their leaves maximize their surface area in order to absorb light, but this is a liability in strong winds. John Smeaton's Eddystone lighthouse (1759) used the tree's method of resisting these forces, with a broad bole that tapers towards the top in a pattern that reflects the stresses induced by the wind.

Another early example is Joseph Paxton's Crystal Palace (1851). Paxton was well placed to be bio-inspired because he was first a horticulturist, the most famous of his era, and second a builder of innovative structures. His most famous achievements in the respective spheres were the cultivation of the giant *Victoria regia* water lily, which he brought to flower for the first time in England in 1846, and the Crystal Palace, the iron and glass building erected in record time to house the 1851 Great Exhibition. The giant lily has ribs on its 1.5–1.8 m diameter leaves that appear to have been engineered. The Crystal Palace used iron ribwork to support the glass.

The *Victoria regia* water lily was so large Paxton had to build a new greenhouse for it. In 1850, he gave a talk to the Royal Society of Arts at which he demonstrated the lily's leaf; he said:

> You will observe that nature was the engineer in this case. If you will examine this, and compare it with the drawings and models, you will see that nature has provided it with longitudinal and transverse girders and supports, on the same principle that I, borrowing from it, have adopted in this building.

'This building' was not the Crystal Palace but the glasshouse built at Chatsworth to house the water lily but it is clear that Paxton, without slavishly following the rib structure of the plant, had thought hard about leaf ribbing before building his novel glasshouses. This was the start of a new way of building. Every steel and glass building built since derives ultimately from the Crystal Palace.

For the most part, though, until the 20th century nature and the human builder and architect went about their work in very different ways. As D'Arcy Thompson pointed out, nature has striven to make structures by the least energy principle, using the minimum amounts of materials. Some of her vast range of solutions include the skeletons of sea creatures such as the radiolarians, the hexagonal structures of wasp nests and bee honeycombs, and eggshells, where there is no skeleton or scaffolding, the whole structure being a single curved surface.

But for most of history, human building has employed simple stone-on-stone and brick-on-brick techniques. It was the particular demands of bridge-building that broke this stranglehold. The most intuitive way of spanning any space is the horizontal beam, the lintel. But an examination of the remains of classical temples, such as the Parthenon, will reveal that many of the lintels are cracked. Only a short space can be spanned in this way before the beam cracks under its own weight. This is because in the world of gravity everything hung out to dry sags in the middle. If you stretch a rope or cable between two posts, however hard you pull there will always be a sag in the middle. If the object is a beam, the sag means that although the top of the beam is being compressed, the bottom surface is being stretched (it is in tension). Bricks are so strong when they are compressed that the Tower of Babel could have been built almost 2 km high before the bottom bricks would have started to crumble under the weight above them. But bricks can be pulled apart (in tension) very easily. That is why bricks can be split easily by cracking them with a blow from a trowel or a karate chop – a violent blow to one side of a brick creates high tension in the opposite face and this tears the brick apart. Similarly, a stone or concrete lintel, if asked to span too great a distance, will simply crack in the middle.

This means that a new strategy for building brick or stonework spans needed to be devised. The arch is the best solution. In an arch, each brick pushes sideways against its neighbour and the whole structure is in compression. The weight of the bridge is directed along the line of the arch down to earth. Once an arch is complete, collapse is virtually impossible unless the side walls move outwards – they sometimes do but that is a fault in the structure of the abutments, the structural supports designed to take the sideways thrust of an arch.

It was Isambard Kingdom Brunel who showed what can be done with arches on a grand scale. When the Great Western Railway had to cross the Thames at Maidenhead he designed a bridge with the flattest arch in the world. Many predicted that it would fail but it is still standing over 150 years later.

But there is a limit to what can be done with such techniques and the way forward was with a different type of structure altogether. Behind all buildings lie two simple principles: pushing (compression) and pulling (tension): any point in a structure is either in tension or compression – it cannot be doing both at once. In the 20th century, building moved decisively towards using tension structures – as nature does – rather than compression.

In order to build a successful bridge two things are required: 1) a perfect understanding of the forces of compression and tension; 2) adequate materials. Materials were a problem for a long time. Bridge-builders could not go beyond the arch until a material strong enough in tension was available. And that material was steel.

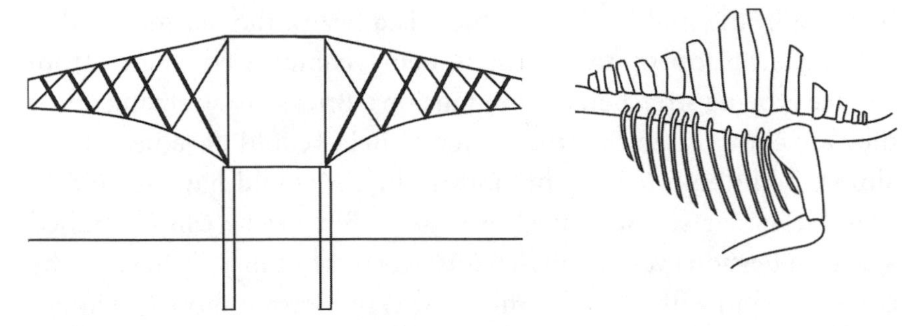

Fig. 9.1 The structures of the Forth Bridge and the forequarters of a large quadruped, as compared by D'Arcy Thompson. The bolder lines in the bridge are rods in compression, the fainter lines are in tension. The resemblance is most noticeable in the skeleton of the bison: in the living animal the tension members would be the ligaments.

Cast iron was no good. This, like the stone it replaced, is strong in compression but weak in tension. The first cast-iron bridge, at Coalbrookdale (1779), was an arch bridge so, technically, little progress had been made despite the novelty of its material. But by the time the Forth Bridge was built, in 1890, steel was available. Steel is the first choice material of the modern world because it is strong in compression *and* tension. Ultimately, steel's tensile properties would lead to new lighter, longer-spanned bridges.

The Forth Bridge is a cantilever bridge, a structure engineers share with nature: the backbones of quadrupeds are cantilevers, as D'Arcy Thompson pointed out (fig. 9.1). The principle is that to launch a horizontal deck from a pier there must be a balancing weight on the opposite side. It is the same principle seen in large horizontal jib cranes – to balance the slender jib projecting many feet from the tower, adjustable concrete weights are provided on the opposite side. Generally, cantilever bridges resemble quadrupeds in having two cantilevers joined in the middle. In horses and cattle, to balance the neck and head, the front cantilever is much larger, heavier and stronger than the back; in the Forth Bridge, the weight is equally balanced around the massive piers with only a light joining section where the cantilever reaches the limit of its stability and butts up against the next section.

The Forth Bridge embodies another natural principle besides the cantilever. The main diagonal struts are steel tubes over 4 m in diameter – and steel tubes, like the stems of plants, are prone to buckling. The tubes of the bridge are stiffened by six T-pieces running the length of the tube and many plant stems such as bamboo similarly have six bundles of fibres running along the length.

For thousands of years human beings built only with compression. A giant cantilever like the Forth Bridge has some parts in compression and some in tension, but tension structures really entered the builder's lexicon with suspension bridges.

The essence of a suspension bridge is to span the complete distance with two hanging cables (chains in the early versions) and to suspend the deck from these cables with smaller vertical cables. The piers need to be tied back very strongly into the abutments to balance the huge tension in the cables supporting the deck and it is the piers that take the compressive force. Although arch bridges can be very beautiful, suspension bridges brought a new kind of structural grandeur into the world, one that seemed to epitomize the age. The great suspension bridges – Brooklyn (1869), Sydney Harbour (1932), the Golden Gate (1937) – became famous and Brooklyn Bridge especially attracted poets. The Russian futurist Mayakovsky imagined:

> … enough hands, enough grip
> while standing,
> with one steel leg
> in Manhattan
> to drag
> toward yourself
> Brooklyn by the lip!

Hart Crane's 1933 epic of the United States takes Brooklyn Bridge for its title (*The Bridge*) and opening invocation: 'O harp and altar of the fury, fused,/(How could mere toil align thy choiring strings).' 'Choiring strings' is apt because suspension bridges resemble lyres or harps and they bring music to the landscape.

If the tensile cables of suspension bridges brought tension and curves into architecture the other great spur to organic design was reinforced concrete. Reinforced concrete was invented by a Frenchman, Joseph Monier, in 1867. It was the first of the composite materials that are such a feature of bio-inspired engineering. The principle is satisfyingly harmonious. Concrete and steel are complementary – concrete is strong in compression, very weak in tension, steel strong in both but more expensive than concrete. A steel grid inside a concrete beam takes the tension and puts the concrete under some degree of compressive strain. A later development, pre-stressed concrete, patented by another Frenchman, Eugene Freysinnet, in 1928, takes this further. If the steel within the concrete is put under considerable tension (is pulled at the ends), and the concrete allowed to set around it, when the beam is dry the steel will still be under tension, thus forcing all of the concrete into compression. When, in use, a tensile stress is then applied to the beam, the concrete will remain in compression until a tensile force greater than the pre-stressing is applied. This enables almost any size or shape to be cast in concrete.

The potential of reinforced concrete for spanning was realized by the great Swiss bridge-builder Robert Maillart, an alumnus of the Swiss Technical University in Zurich, which has a record over a century of producing major engineers and architects. And while reinforced concrete was a fine medium for bridge-building, it could span anything. The Spaniards Edward Torroja and Felix Candela (who lived and worked in Mexico for most of his life), the Italian Pier Luigi Nervi, and the Brazilian Oscar Niemeyer all made memorable buildings in concrete. Concrete allowed architects to create far more organic forms than they had with compressive stone and brick structures. The fact that it used the tensile element of the steel reinforcement meant that the limits of normal compressive structures no longer applied. Although there is a stereotype that concrete modernist architecture was anti-nature, often in its use of exuberant curves it was flamboyantly *in love* with nature. This was particularly the case with Niemeyer.

The 1950s was the age of reinforced concrete. It allowed architects and engineers to exploit one of nature's most organic structures: the shell. The great specialist in these structures is the Swiss engineer Heinz Isler. Since the 1950s, Isler has built more than 1,500 delicate shell buildings from very thin reinforced concrete.

More than any other recent engineer or architect, Isler has used natural forces to find his forms. How he came to do this is a fascinating case study. Isler was drawn equally to a love of art, nature and engineering and, given that education usually keeps these disciplines apart, his path was never going to be straight or simple. He had intended to study art but this was 1945 and four months of military service, during which he erected military structures, turned him towards engineering. Like, Maillart, he studied at the Swiss Technical University in Zurich.

In 1954, Isler was a 28-year-old engineer with his first big job: he was working his passage back to art college but the job was challenging: to construct a new, acoustically satisfactory ceiling for a dance hall. The ceiling would need to curve gently towards the walls: how could he find the shapes, the equations to do this? A curve is defined by its radius but these curves would need to vary smoothly in radius, from almost infinity at the centre of the ceiling to a small radius where the ceiling meets the wall. To do the job traditionally, on a standard drawing board, would have required 400 coordinates.

This was beyond Isler's capacity, so he constructed a curved, wooden drawing board, but this only pushed the problem one stage on: how to create the curved shape of the board? It had to be done freehand and this offended Isler's aesthetic sense because only in the hands of a Picasso do freehand curves have the inevitability and perfect gradations of nature's curves. Isler explains how the problem was resolved:

> Coming home at 4 o'clock in the morning, having defined a few new curves of the shape, wanting to sink exhausted into my bed, I was struck by the sight of the pillow on my bed. That was exactly what I had been looking for for so many weeks in vain.

He meant that, unlike his freehand curves, the pillow's curves were dictated by the pressure of the feathers pushing against the tension of the fabric. The resolution of these two created perfect, smoothly varying curves.

To turn the pillow idea to practical use he created an experiment.

He clamped a rubber membrane with grid lines painted on it over a wooden frame. The membrane was pumped up with air till it resembled a pillow. The grid lines followed the surface and defined the curves of the structure.

Isler realized that such models would solve his problem in the most elegant way possible. In the most efficient structures, at every point the strength, stiffness and weight are just what is needed to counter the forces that act on them, no more, no less. And a model such as Isler's balloon membrane takes the shape it does as a result of automatically resolving those forces. It is a sort of divine shortcut to the perfect structure. He had done what Bob Full says you must do: abstract nature's principles. Isler says:

> For the hotel shell it was too late but for future designs I had found a method for forming harmonious shapes. As the inflated shape is rather similar to a soap bubble it proved to have excellent physical properties such as minimal surface, minimal weight, minimal material, equal medium curvature at every point thanks to the equal resistance to bunching at every point, equal stress at every point as well as direction: in short, a unique accumulation of positive qualities.

By this method, 'a whole world of excellent shapes presented themselves' (fig. 9.2). Isler developed a repertoire of form-finding experiments. One of the most productive was a kind of inverse of the balloon experiment: a wet cloth is hung from four points and left out on a freezing night (obviously, only a northern engineer was ever likely to come up with such a technique). It takes up a perfect curve under gravity and freezes into that curve. Inverted, it is a model of a perfect shell roof.

Isler's shells, the results of these experiments, are human constructs that fulfil many of the criteria of natural objects. This puts them in the mainstream of bio-inspiration even though Isler is largely unaware of the tradition. When Isler says that at the Hotel Kreuz he was seeking curves that would 'vary smoothly in radius' it was the same principle that D'Arcy Thompson had expounded 37 years before in *On Growth and Form*. And it is significant that Isler invokes the soap bubbles to which D'Arcy Thompson continually

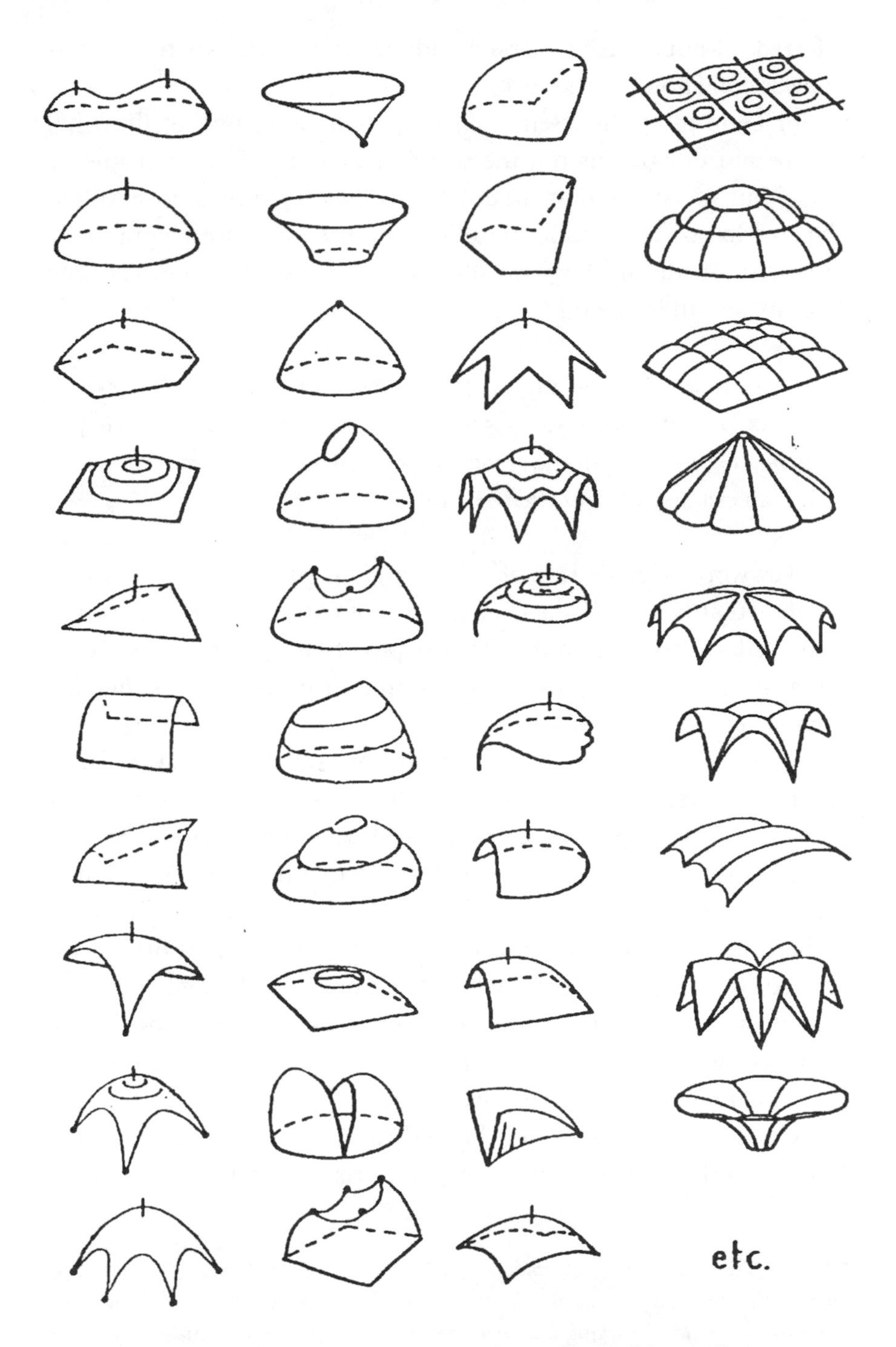

Fig. 9.2 Heinz Isler's gallery of sketches of possible shell structures resembles an inventory of biological forms – jellyfish perhaps.

referred, although Isler worked independently and to this day is unaware of Thompson's work.

For Thompson, the essence of beauty of line in painting, drawing, nature and design was not the simple geometry of straight line or circle, but of curves that modulate from one gradient to another. This is what we respond to in calligraphy, the way a stroke gradually broadens or diminishes; we also see it in sand dunes. D'Arcy Thompson put it like this:

> The Florence flask or any other handiwork of the glassblower is always beautiful because its graded contours are, as in its living analogues, a picture of the graded forces by which it was conformed. It is an example of mathematical beauty …

The key word is 'graded'. In glassblowing, the pressure of the air acts on glass that has gradients of temperature within it: the curves that result reflect these gradients.* The shapes of Heinz Isler's shells are a literal realization of D'Arcy Thompson's idea: *the curves follow the contours of the lines of stress.*

In *On Growth and Form*, Thompson relates how the Swiss engineer Professor Karl Culmann, at the Swiss Technical University in Zurich, was designing a crane in 1866 when he happened to visit the dissecting room of the anatomist Hermann Meyer. Meyer had sectioned a human femur, revealing its internal structure. Culmann realized instantly that the fibres were taking up the lines of tension and compression produced when the bone was loaded with the weight of the body (fig. 9.3). 'That's my crane!' he exclaimed. The engineering structures that emerged from the tradition established by Culmann would similarly advertise their lines of force. Culmann's work was a direct inspiration behind the form of the Eiffel Tower (1889). The broad base of the tower converts the tension in the upper

* D'Arcy Thompson's idea of graded curves echoed William Hogarth's attempt to analyse the geometry of beauty in the 18th century. Hogarth wrote: 'A graded lessening is a kind of varying that gives beauty. The pyramid diminishing from its basis to its point, and the scroll or voluta gradually lessening to its centre, are beautiful forms.'

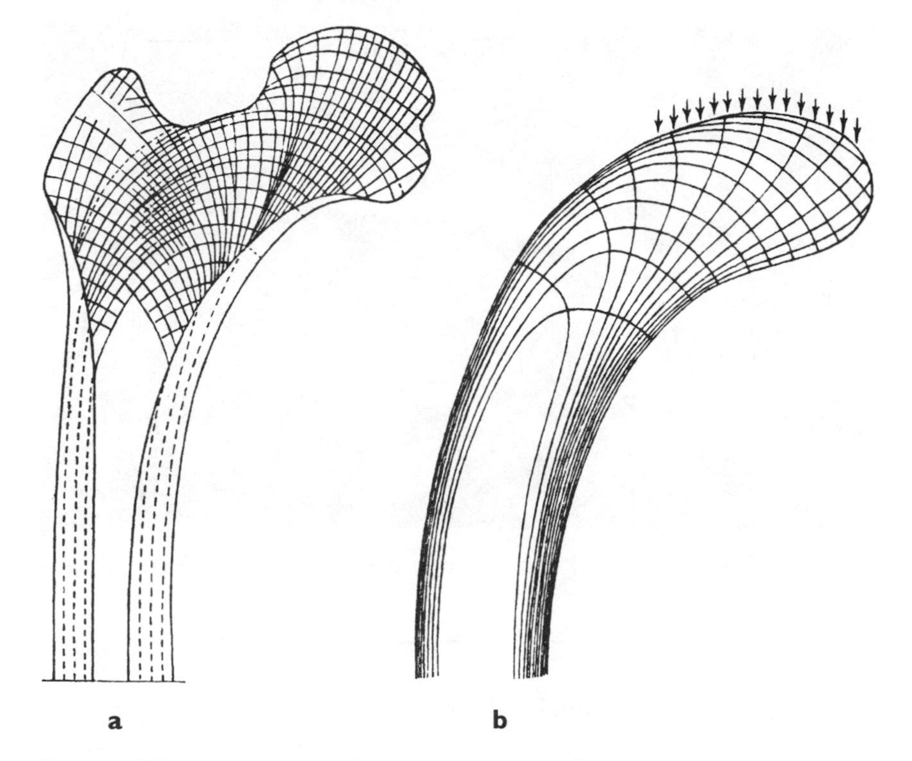

Fig. 9.3 'That's my crane!' the Swiss engineer Karl Culmann exclaimed in 1866 when he saw a section through the femur. The internal structure of the femur (a) follows the lines of stress and Culmann's crane head (b) took a similar form.

---

tower caused by the wind into compression.

Culmann was an inspirational figure at the beginning of a great tradition of Swiss engineering that includes Isler, Maillart and, more recently, Santiago Calatrava (Calatrava is Spanish but he studied at the Zurich Technical University and is based principally in Zurich). Harry Lime's notorious quip in *The Third Man* that 'In Switzerland they had brotherly love, five hundred years of democracy and peace, and what did that produce … ? The cuckoo clock' is belied by the country's engineering tradition, centred on this institution. This excellence has been attributed to the country's hybrid French and German culture. Marrying French aesthetics and German technological rigour, the Swiss have produced wonders of form and function in harmony.

This is not to claim that bio-inspired architecture will always

Fig. 9.4 Heinz Isler's shells over a filling station on the Berne–Zurich highway (1968).

automatically be beautiful, but there is no doubt that the forces that form living structures, and that also act in bio-inspired materials science and engineering, produce the kind of graded curves that Hogarth and D'Arcy Thompson agreed constituted beauty.

Isler designs his shells by scaling up from his models, and the standards of construction are exacting. They have a delicacy that totally confounds concrete's stock image (fig. 9.4). Classical domes have a thickness/radius ratio of about 1:50; natural shells such as eggshells come in at 1:100; some of Isler's domes have a ratio of 1:800. Such shells need no painting and the surface, if properly made, smooth and everywhere in compression, will never leak.

Isler is so wedded to his naturally-derived forms that when, eventually, computers were able to create similar structures, he felt angry. His biographer, John Chilton, Professor at Lincoln School of Architecture, reported his reaction like this:

> He said that it was as if he had been climbing a wall, a high wall in the mountains, a very dangerous and lonely path, but he climbed it, and when he finally reached the top he found a valley which nobody had ever seen before. There was a paradise, full of flowers, rare or

> unknown, and in these he lived and could make his forms. He thought there might well be others who would follow him, climbing up by the same or [a] similar way. Then suddenly one morning there was a loud noise and out of the valley side there came a tunnel with a road, a motorway, where others could come to his valley in complete comfort and with little effort. He felt robbed and that everything had been taken away from him.

Despite this, computer simulations have powerfully confirmed the structural principles behind Isler's shells. Although he has always worked as an engineer and does not call himself an architect, Isler is, above all, an artist. This is apparent both from his beautiful structures and the winter-wonderland ice sculptures that he creates in his garden. Rather like the English environmental artist Andy Goldsworthy, he sprays water over structures in the garden, both natural and man-made, to create fabulous ice grottoes. In a talk given to the International Association of Space Structures (IASS) Structural Morphology group in Nottingham, in 1997, Isler said of these works:

> When on a cold winter night – well protected by a warm, weathertight overall – you begin to experiment with the gardener's hose, you really enter a new world. Your wildest imagination is modest compared with the richness nature can produce.

Unlike his shells, which are built to last, these structures remain only in photographs. Of his 1,500-plus domes worldwide, only one site in Britain can boast an example: the Norwich Sports Halls built in 1987–91. The domes were built because the English architect Tony Copeland worked in Switzerland for 25 years before returning to England to work near Norwich. In Switzerland, he came to respect the European tradition of building with reinforced and pre-stressed concrete: 'Knowledge of concrete technology is different in Germany and Switzerland; Britain is a steel frame and light structures country.' Copeland worked on some 30 shells with Isler in Europe before proposing the Norwich project. He is a great believer in Isler's shell structures:

> It's like building an egg – the bottom part is the four foundation piles

> and their pre-stressed cables. If you build on bad land, near lakes for example, even if the piers move the shell can roll without cracking. One was erected on top of a supermarket once – there were problems with the building and the only part that was 100 per cent was the shell.

Concrete domes have become unfashionable, and their niche has been usurped to some extent by membrane structures such as the Millennium Dome. Even a building like the US Air Force Pavilion at Duxford Aircraft Museum, designed by Norman Foster, which looks like a shell structure, is not constructed on the Isler principle. But, as Tony Copeland points out: 'Shells are perfect for swimming pools: you don't need the volume, there's less to air-condition; conditions are perfect inside, the acoustics are very good.' Isler's work remains a shining example of the possibility of harmonious collaboration between engineers and the forces of nature. David Campbell, CEO of one of the most creative structural engineering companies, Geiger Engineers, says of Isler's shells: 'They're gorgeous, absolutely gorgeous. Calatrava does architectural shells, like Gaudi; Isler does engineering shells. They're pure expressions of functionality. There's aesthetics involved but he's looking to do the most with the least. The architecture comes out of that.'

As reinforced concrete fell from favour in the 1970s, tarnished by the failures of system-built flats and the excesses of Brutalist architecture, tension structures spread from bridges into almost everything. And the dominant technique of contemporary bridge-building is not the suspension bridge but cable-hanging.

Many people, glancing idly at a modern cable-stayed bridge such as the Second Severn Bridge, assume it is a suspension bridge. But in such a cable-hung bridge there is no continuous cable running the whole span: each cable is tied down to the deck. The tension created by the weight of the bridge on the cables has to be balanced: this is usually done by cables on the other side, so that the horizontal forces on the pier cancel out. Cable-hanging also combines very well with metal arches to create hybrid structures. Technically, cable-hung bridges have slightly less reach than suspension bridges but they are cheaper, using less material.

One figure brings together the two traditions of reinforced

Fig. 9.5 Santiago Calatrava's *brise soleil* for the Milwaukee Art Museum (2001). The *brise soleil* is retractable.

concrete and cable-hung tensile structures: Santiago Calatrava is currently the leading exponent of organic architecture. His 2004 Athens Olympic Stadium has made him one of the half dozen best-known architects in the world today. Calatrava trained in both architecture (at Valencia, where he absorbed the Spanish tradition) and civil engineering (at the Swiss Technical University in Zurich). His most famous bridge is probably the Alamillo Bridge built in Seville for Expo '92. The Alamillo Bridge introduces a playful variation on cable-staying: instead of having cables on both sides of the piers, there are only cables on the decking side. The pylon leans backwards at 58° so that gravity does the work of counter-balancing the tension in the cables. The result is that the bridge looks like a giant harp: a witty, metaphorically expressive gesture. Some engineers have objected to Calatrava's bridge because it flouts the minimalist principle: for visual reasons, the pylon is larger than it need be in strict engineering terms. But this seems a churlish response to such a joyful structure. Calatrava's mentor Felix Candela has said of his work that it has 'that simplicity, delight and cheerfulness in appearance that Michelangelo regarded as the essence of a true work of art'.

Calatrava's organicism is ever apparent. The poster image for the *Zoomorphic* exhibition was his Milwaukee Art Museum (2001), perhaps the largest biomorphic building in the world. Milwaukee is not entirely Calatrava's structure: the original museum was built by Eero Saarinen in 1957 and Calatrava added to this the giant signature *brise soleil*, amongst other features (fig. 9.5). The *brise soleil* has been likened equally to a bird's wings and a whale's tail. You take your choice: being more enamoured of whales than birds I incline to the latter.

Calatrava first made his name with transport termini, beginning with the Zurich rail station. The Lyon airport and rail station is one of the most dramatic and organic with a great sci-fi beaked monster digging its snout into the ground. The roof of the Lisbon terminal evokes forest tree canopies and the two radiating steel grids of the entrance are surely the antennae of a well-endowed moth. Calatrava is currently designing the new subway station for the World Trade Center site.

Calatrava is both engineer and architect and reactions to his work comprise a case study in the fault lines between these two disciplines. Calatrava is also a sculptor and his buildings, however much of the engineer he puts into them, are sculptures. David Campbell sees his work as 'modern Gothic'. Certainly the Lisbon train-shed roof fits that description and the elements of gratuitous grandeur that you find in buildings such as the Milwaukee Art Museum seem Gothic to an engineer. The Olympic Stadium is more functional and reminds us that if brick arches were the dominant spans of the pre-steel era, it is steel arches that are mostly used to keep all those tensile cables aloft today.

Cable-hanging is not the last word in making beautiful bridges – it is still possible to devise totally novel bridges. A project in direct line from D'Arcy Thompson's work is the Dinosaur Bridge (fig. 9.6) designed by David Marks and Julia Barfield, architects of the London Eye (itself a notable tension structure: like a gigantic bicycle wheel).

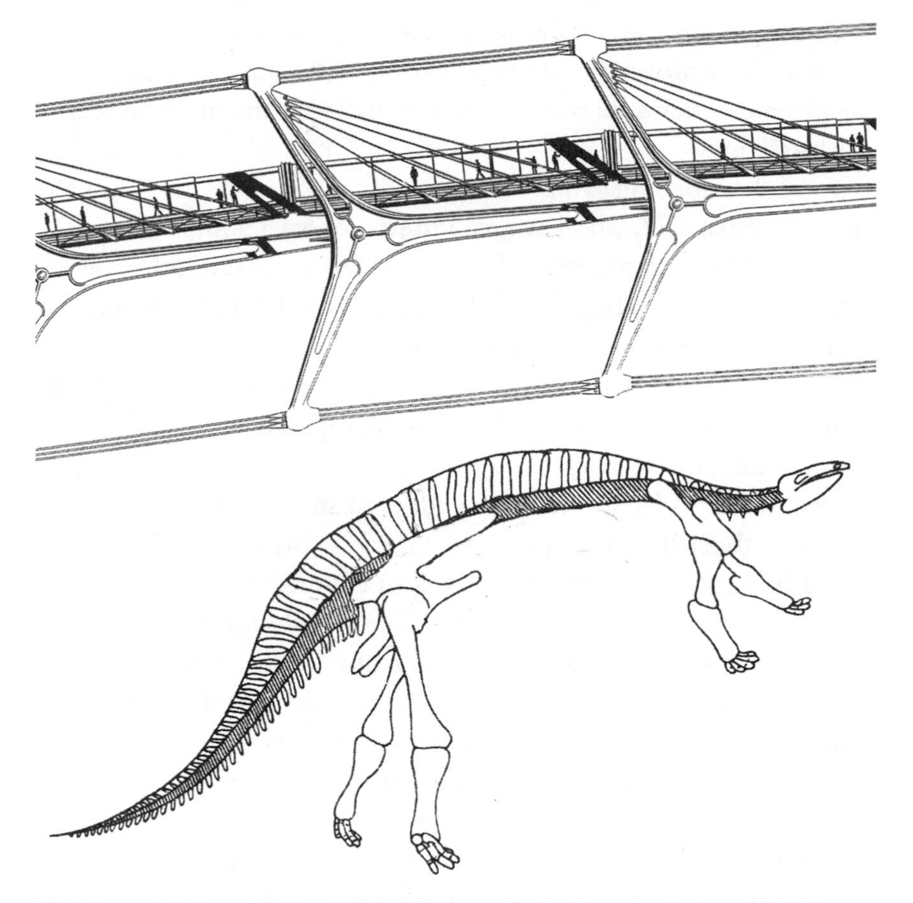

Fig. 9.6 Marks and Barfield's Bridge of the Future design (1988) was explicitly inspired by D'Arcy Thompson's account of the dinosaur spine.

The Dinosaur Bridge won a 1988 competition organized by *New Civil Engineer* for 'The Bridge of the Future'. The bridge is as yet unrealized although it is a perfectly feasible engineering structure (it is a favourite project for student research groups to check out its functional viability). Unlike mammals such as horses and cattle, in a dinosaur like the *Stegosaurus* it was the rear cantilever that was the strongest. The enormous tail balanced the body and long neck around the huge hindquarters, while the forelegs were small – the animal could easily rear up and balance around its rear cantilever. D'Arcy Thompson said:

It is easy to see that, just as the quadruped mammal may carry the

> greater part but not all of its weight upon the forelimbs, so a heavy-tailed reptile may carry the greater part upon its hindlimbs, without this process going so far as to relieve its forelimbs of all weight whatsoever.

Marks and Barfield, who have read their D'Arcy Thompson, took it one step further: they designed their bridge so that it *does* take all of its weight on the hind cantilever. This is anchored in concrete and the bridge snakes out like the dinosaur's spine to land lightly on the other bank, where it can slide in and out as the loading on the bridge varies. The 'vertebrae' are pulled into compression by the tensile cables linking them.

Ideally, the bridge would need a gorge to show it off to maximum effect and Julia Barfield has jokily suggested the Grand Canyon. It would bring the question of nature and engineering to a satisfying full circle: a modern structure based on a dinosaur extinct for 65 million years spanning rocks that date from 1.7 billion years ago.

There is another approach to structures which has strong roots in nature: it began with Buckminster Fuller. His geodesic domes, created from complex webs of triangular or hexagonal and pentagonal units, have found many echoes in natural structures. Fuller had been developing his distinctive structural geometry since the 1920s. His aim was to produce economical solutions to building problems and he waged a lifelong war against the tyranny of the right angle, believing that the cube is not the best shape for buildings (some of his students used to joke that 'Bucky probably invented the triangle'). But building large spheres creates special problems and Fuller devoted great ingenuity to this question (fig. 9.7). Fuller's geodesic dome was patented in 1954 and thousands of domes have been erected all over the world, the most famous of which was the American pavilion at Expo '67 in Montreal.

Fuller's search for spherical grids that use the minimum amount of material was prefigured by nature's search for the same during evolution. The icosahedron is a close relative of a sphere with flat triangular sides. It is found in nature in the bacteriophage, which we considered for its self-assembling properties (*see* page 143). The biologist Aaron Klug discovered the icosahedral structure of phages

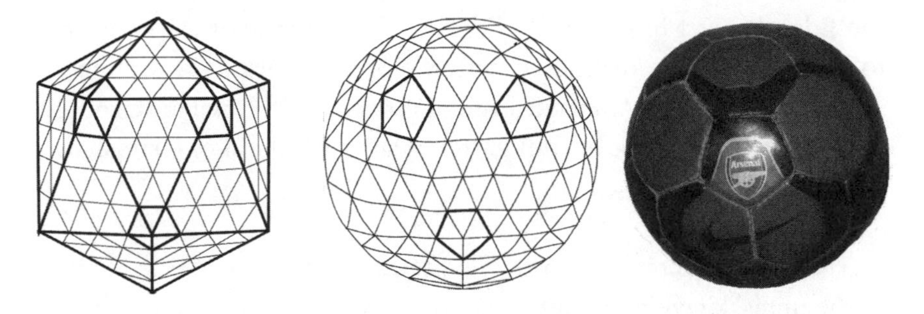

Fig. 9.7 Nature and Buckminster Fuller both wrestled with the problem of how to create spheres, or almost spheres, from rod-like components. Fuller began with the 20-sided figure the icosahedron – a form also favoured by nature – and divided it up into smaller triangles. The resulting structure looks as if it has enough triangles to be buildable. To turn it into a sphere, knock down the sharp corners and tweak out the middle of each large triangular face. The result will be a sphere with slightly distorted triangles.

The result also shows another pattern. Where the corners of the icosahedron were there are now pentagons; everywhere else the figures are hexagons. So to create a sphere you can dispense with the triangles and use hexagons and pentagons. This is the classic structure found time and again in nature and human culture: from radiolarians to footballs and from the C_{60} molecule buckminsterfullerene to geodesic domes.

---

in 1962 and drew Fuller's attention to it.

When nature wants to create a sphere using the minimum of materials she uses the hexagonal/pentagonal version of the geodesic dome. The versions found in nature are tiny but the patterns are the same. A geodesic dome on an even smaller scale was discovered in 1985 by Harry Kroto and his team at the University of Sussex: a new form of carbon, a spherical molecule with 60 atoms. It is a classic hexagonal-pentagonal geodesic dome and was named buckminsterfullerene or the 'buckyball' in recognition of this. Buckminsterfullerene is a molecule at the bottom end of the nanoscale, with a diameter of 2 nm. The hexagonal/pentagonal sphere is one of nature's prime structures. It is found in every size: from the nano (C_{60}: buckminsterfullerene), through the micro (radiolarians – *see* page 153), the everyday scale of human objects (the football), to the architectural (Fuller's domes) (fig. 9.7).

One of the most dramatic recent uses of the geodesic dome is at the Eden Project in Cornwall. Because the purpose of the project is to focus awareness on the ecosystems of the world, the architects, Anthony Grimshaw Associates and engineers, Anthony Hunt, sought architectural inspiration from nature. David Kirkland of Grimshaw's said of the project:

> The best structures, demonstrating beauty and function, are built by the animal kingdom. Termites' and wasps' nests have air conditioning, and they have farms. Nature takes what is available and produces a natural solution.

The Eden domes are recognizably geodesic but they have some specific idiosyncrasies. The external appearance is dominated by the very large hexagons (with the usual pentagons to achieve closure) and the cladding is air-filled ETFE, a more or less transparent close cousin of Teflon. These give it the appearance of a gigantic bubble-wrap. Also, unlike most geodesics, these domes intersect like soap bubbles. The bubble analogy is not just a visual simile because to make the joint between two domes, the properties of bubbles were exploited. Barry Johnson, for the contractors McAlpine, explained:

> It's a natural fact that where bubbles intersect they do so vertically in a plane, which means that you can put an arch between the two bubbles and it will automatically be vertical and straight.

The Eden Project is a triumph of organic design – the original site was a china-clay pit and the domes sit well inside the pit. Which makes them truly look like natural structures, extruded from the earth, like giant fungi that have sprouted overnight.

An interesting area where biology meets large structures is tensegrity. Tensegrity is one of Buckminster Fuller's fancy words, although he did not invent the concept. The word is apt because it combines 'tension' and 'integrity': a tensegrity structure is one which is held together by tension members but which maintains its shape. This is paradoxical because tension members – wires and rubber bands – have to pull on something to maintain their shape: you cannot make a structure out of rubber bands alone. So in any structure involving tension members there have to be solid structures, such as rods, to prevent the wires collapsing in a heap of spaghetti. The innovation of tensegrity structures is to allow the tension members to maintain their shape by pulling on stiff rods that *do not touch each other*. So the external shape of the structures consists of members entirely in tension: indeed they can be made with one continuous wire.

Such structures are far easier to grasp as an object than to describe, and simple tensegrity models can be made from drinking straws and rubber bands. Like Miura's map, tensegrity lends itself to kitchen-table experimentation. For the simplest kind you need three plastic drinking straws, six paperclips and nine rubber bands (*see* fig. 9.8). The paperclips should be snapped in two so that they have a single loop that can be threaded into the straws. The unstretched rubber bands should be about two-thirds the length of the drinking straws. It is difficult to specify a logical order for connecting the pieces. Each straw end has three rubber bands connecting it and from one aspect the bands form a triangle and from the other a parallelogram. It is best to start with one triangular end. When the structure is all wired up you can see that none of the rods actually touch each other: they're held in equilibrium by the rubber bands. If you look along the triangles and try to line both ends up you will see that the triangles are displaced: one triangle is rotated against the other; similarly with the parallelograms.

Even this simplest model has very interesting properties. Although drinking straws are very weak, with a tendency to buckle, the tension bands hold them in such a way that the compressive force is always directed straight down the tube and buckling doesn't happen. It is immensely fiddly to assemble – pieces keep falling apart – but once the last band is secured you can fling the object around, squash it and it seems to be almost indestructible.

The structure is not symmetrical in its properties. In one direction it squashes flat and bounces back. In the other direction, it resists the pressure like a shock absorber. If you were God, wanting to create versatile three-dimensional structures out of nothing very much, tensegrity would take some beating. But what does such a structure have to do with biology? For Donald Ingber, a professor at the

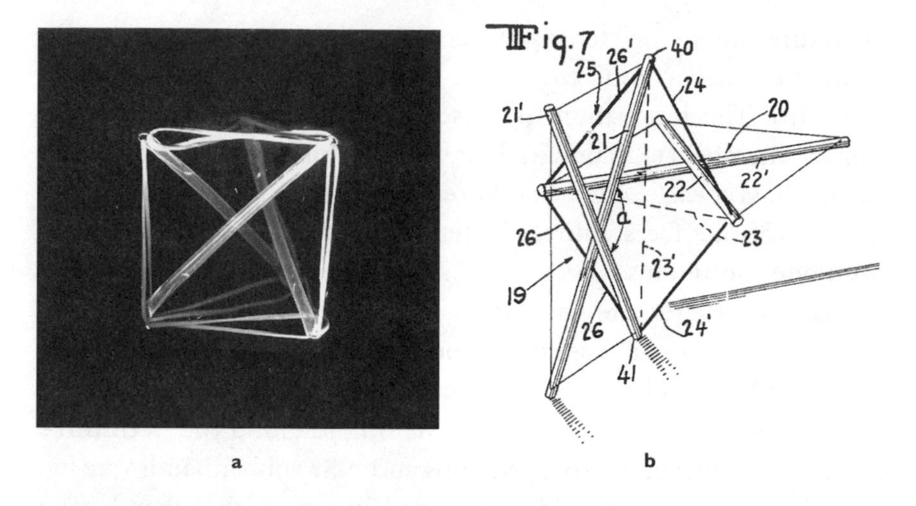

Fig. 9.8 a) A simple home-made tensegrity made from drinking straws, rubber bands and paperclips; b) a diagram from Kenneth Snelson's patent (US Patent 3,169,611).

Harvard Medical School, 'our bodies provide a familiar example of a pre-stressed tensegrity structure: our bones act like struts to resist the pull of tensile muscles, tendons and ligaments, and the shape stability (stiffness) of our bodies depends on the tone (pre-stress) in our muscles'.

If you look at a skeleton, and note, in particular, the absence of anything between the pelvis and the ribcage other than the lumbar vertebrae, it becomes clear that we are held together by tensile members. By themselves, muscles would be no better than rubber bands, but the rigid bones give the muscles something to pull against. Ingber goes further and claims that the shape of living cells is maintained by tensegrity structures within the cell.

Although the human skeleton, with its accompanying musculature, is such a suggestive example of tensegrity, the idea did not emerge from biology. In fact, its genesis was muddled. In 1996, the *International Journal of Space Structures* carried pieces by the main protagonists, the sculptor Kenneth Snelson, Buckminster Fuller and the French engineer David George Emmerich, in an attempt to clarify priority. Emmerich seems to have discovered tensegrity independently, as did Darwin and Wallace with evolution, but the

Snelson/Fuller story is a fascinating tale of contrasts.

In 1948, Kenneth Snelson was a starry-eyed second-year student at the University of Oregon, a GI student who had already served in the war. In the heady days of peace, he had just cottoned on to the German Bauhaus, when he discovered that one of the Bauhaus artists, Joseph Albers, was in residence at Black Mountain College, a hotbed of modernism and – to the local population – a den of communist iniquity. He decided to go to the summer school to study with Albers and others, including Willem de Kooning and John Cage.

Buckminster Fuller was a late addition, a substitute, who arrived two weeks into the session. Snelson had no particular prior interest in Fuller's work, but came under his spell:

> In those days that seem so long ago, just after World War II, the country was upbeat and Bucky was an attractive optimist who made his audience believe that the world would be okay if only things were done his way. He had the right stuff for a true cult leader: charisma and a message. The enemies were out there to defeat.

Under Fuller's influence, Snelson drifted from painting into sculpture; he tried to cross the ideas of Albers and Fuller and in the fall/winter of 1948 made some mobiles of the kind invented by Alexander Calder, and now found hanging in millions of children's bedrooms.

Then, instead of just hanging things from wires, as in the conventional Calder mobile, Snelson thought that he would try to balance some of the structure on the wires. He wanted to create an air of marionette illusion by having part of the structure float on almost invisible wires. The resulting *X Piece* no longer moved like a mobile but it certainly had an aura of mystery.

Snelson returned to Black Mountain in 1949 and showed the *X Piece* to Fuller, who looked at it for a long time before saying, 'Ken, may I keep this?' Now, unlike Snelson, who was groping in the dark, Fuller had been feeling his way towards a new form of tension/compression structure for some time. Fuller claimed credit for the idea and for many years Snelson felt unable to press his claim. But in September 1959, Fuller's assistant John Dixon invited Snelson to see

Fuller's show then being mounted at MOMA, New York. A 9 m high tensegrity mast was included in a group of 'Three Structures by Buckminster Fuller'. Snelson had shied away from a showdown: 'The truth was, I was afraid of the man; the enormous father figure. Bucky was not known to abide challenges, and those around him were familiar with his tantrums when he was crossed.' Fuller started to wax lyrical about the tensegrity structure: 'Pure jewelry, Ken, pure jewelry.' Snelson realized he had to speak up now or forever hold his peace: 'I hope my name is going to be on it this time, Bucky.'

Fuller made reassuring noises: 'I'm sure I've told Arthur [Drexler, the head of MOMA] all about you.' John Dixon didn't believe this and introduced Snelson to Drexler. The caption was altered to credit Snelson with the invention of tensegrity sculpture. This was the turning point for Snelson: some of his tensegrity sculptures were included in the exhibition, including a replica of the lost *X Piece*.

Snelson now felt able to reclaim his discovery – no one else had worked with the *X Piece* structure and he decided to patent it (fig. 9.8). Fuller also patented a tensegrity structure. Snelson went on to make many tensegrity sculptures such as the *Dragon* (2000–3) (fig. 9.9).

In the 1980s, architecture began to appear that embodied the tensegrity principle, although the relationship between the building structures and sculptures remains controversial. The key protagonist was the engineer David Geiger, who had already made a name for himself with air-inflated domes and membrane structures in the 1970s. Geiger developed his tensegrity cabledome system in 1984 and patented it in 1988; the first of his tensegrity structures to be built were two domes in Seoul for the Korean Olympics in 1988 (fig. 9.10). David Geiger died in 1989, but in the last year of his life he had built two more cabledomes in America. His firm, Geiger Engineers, continues the work, and cabledomes are increasingly part of the repertoire for roofing large arenas.

David Campbell, now CEO of Geiger Engineers, worked with David Geiger as he developed the cabledome. From the Geiger offices in Suffern, New Jersey, some 35 miles from New York, he explained how Geiger 'started playing around with this idea that became the cabledome. Most of us working with him thought it was a bit harebrained.' Until they ran computer simulations, that is, and discovered

Fig. 9.9 Kenneth Snelson's tensegrity sculpture *Dragon* (2000), Daibiru Building, Osaka, Japan.

that it was surprisingly stable. David Campbell says of the structure: 'I have to be careful in the way I refer to it because in the sense that some of the purists talk about tensegrity, it certainly isn't. It relies on a closed compression ring at the perimeter.'

Snelson's sculptures are bounded only by their continuous tension bands so there is a distinction here between two different, albeit related, kinds of structure. At first, the technique seems to defy gravity. Under high tension, a cable may look straight but there will always be a sag; it is certainly never going to rise in the middle. But

Fig. 9.10 'You walk in and see these floating struts. Everything else is just spider webs' – David Campbell's description of a cabledome. This is the first such dome under construction: the Gymnastics Hall for the Korean Olympics (1988).

in a tensegrity dome the principal load-bearing ridge cables curve across the top of the dome (fig. 9.11). David Campbell says: 'It's one of the things that's very cool about it. Even engineers can go into the building and say "how does it work, what's keeping it up?" It's the bare minimum. You can't get less than that.'

Campbell likes to quote Fuller to describe the effect: '"Compression members floating in a sea of tension" – that's marvellously poetic and well described. That's how you experience it. You walk in and see these floating struts. Everything else is just spider webs.' Which is a nice reminder that it is not only the material of a spider web that is special: the webs themselves are the prototype for tensile net structures. Something like a spider web was built by Frei Otto as the main stadium building for the Olympic Stadium in Munich, 1972.

At first, the reasons for doubts about the cabledome structure were strong. Strictly, something like a cabledome is a mechanism not a structure. That needs a bit of explanation. To be called a structure there has to be rigidity; a mechanism is something that can move. In the simplest case, imagine trying to make the frame for a wall out of four wooden posts nailed together in a rectangle. You may think you've made a structure but there's nothing to keep it square – the whole thing can be pushed over to make a parallelogram. To make it rigid it has to have a diagonal nailed across it – now it can't move: it's a structure.

So there was a worry that it would be unstable: 'But it's not bouncy. That was our concern when we were playing around with it. We were all surprised but David wasn't. As a deformation takes place, it adjusts and it tries to restore itself. Like a membrane. You're deforming its network and the tension in it wants to go back.'

These cabledomes are remarkably light and strong. There have been many studies of their stability and they have withstood whatever the weather has thrown at them since. They do not vibrate if something disturbs them, as David Campbell explains: 'It has very low mass so when you get it excited it doesn't have a lot of energy. It tends to get stiffer as it deforms.'

Kenneth Snelson now works from a ground floor studio in SoHo, New York, with the famous fire-escapes snaking up the outside of the

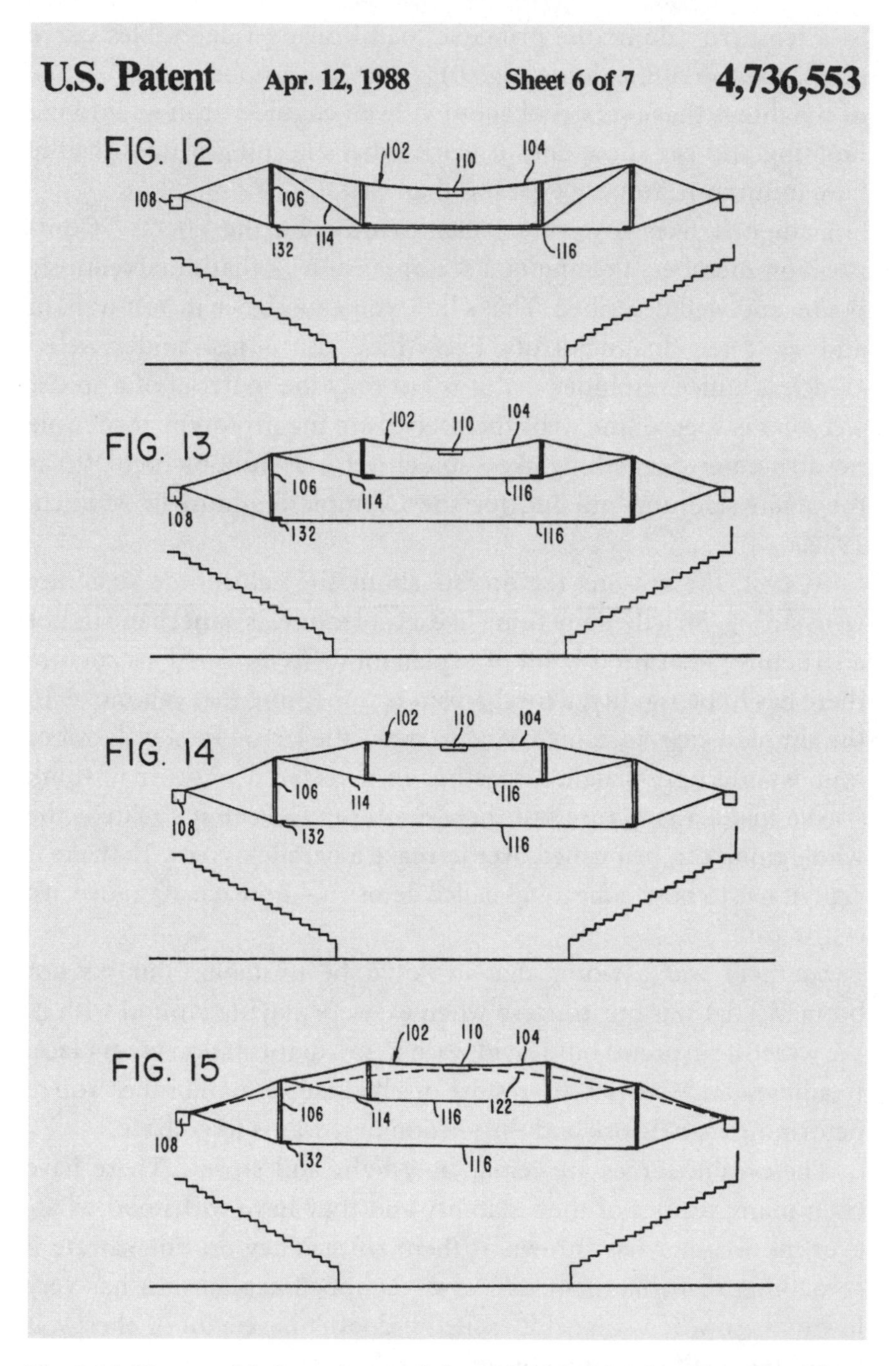

Fig. 9.11 How a cabledome is raised, from David Geiger's patent. Here there are only two sets of cables, one hung within the other, but more layers can be added, raising the dome further.

building and, inside, some fine examples of his sculptures, a few model planes and models of atomic structure. He has a genial, teasing manner and his boyish twinkle belies his 77 years. But he is wryly philosophical about what has been made of his discovery. Snelson is a maverick artist and inventor in the American tradition of sturdy individualism. His combative conversation reminds me of Frank Zappa. Did he think when he discovered it that tensegrity might have an importance beyond sculpture – in biology and architecture? 'The moment I saw it I knew it was all of these things,' he says mock portentously, then collapses into laughter. 'Nonsense! I'm not sure it is, even now. People make claims for it now and Fuller said of it: "Properly examined every structure in the universe is tensegrity." *Properly examined*, what does that mean? It's nonsense. Properly examined, you are me and I am you but we're different people.'

He stresses that just because all things in the world are either in a state of tension or compression that does not make them tensegrity structures: 'A step ladder is not tensegrity. Not all structures are pre-stressed structures, first of all. Tensegrity, if there is any novelty in it, means pre-stressed structures. So where do you begin: is the atom a pre-stressed structure? – I don't think so. It's a stable structure. Certainly, forces are involved in every bit of matter that you have but they resolve in different ways.'

A pre-stress is a stress given to a building component before assembly which then counters forces it experiences in its place of use, for example in pre-stressed concrete, the metal reinforcing rods, stretched before the concrete is cast around them. To make the simple home-made tensegrity structure I described, the rubber bands have to be stretched – that is the pre-stress.

For Snelson, the whole thing was garbled by Fuller: 'Within five years, after I showed it to Fuller, he made up the name tensegrity, then he began to say that not only was this kind of structure to be considered tensegrity but that many things that he had done earlier were tensegrity. So I said: "What's the point of having a new word: if all those things you did back then were tensegrity why have you invented a new word?" The absurdity began there and it continued when he began to announce that all things properly examined are tensegrity. So now when you look on the web you'll find tensegrity types, tensegrity exercises, tensegrity cell biology. I laugh at it because I think that people have gotten a buzzword that they've carried a long way.'

For Snelson, the most contentious aspect of tensegrity is its biological role. For a long time, the mechanical properties of cells were ignored by biologists: the cell was just a sort of elastic bag full of very interesting chemicals. But there has to be an architecture: tissue is tough, resilient stuff that keeps its shape. At the very least, the human body is a good way of getting a fix on how tensegrity works, but Donald Ingber takes tensegrity into the hidden world of the cell. In applying tensegrity to the cell, Ingber assigns the role of the compressive struts to the stiff microtubules and the tensile members are the thinner microfilaments. What evidence is there for this? When living cells are placed on a flexible silicone rubber sheet, the cells move across the surface and as they do they produce tension wrinkles in the rubber. This is exactly what a system in continuous tension would do. Ingber has experimented with tensegrity models and their pre-stress also causes a rubber sheet to pucker.

Snelson has sparred many times with Ingber over tensegrity's biological role: 'We were at a conference together a few years ago in Princeton. He gave his talk and we spoke after and I said: "Donald, if what you're describing is tensegrity what I'm doing is not tensegrity so which of us is doing tensegrity?" I don't know how you can continue to discuss if you can't agree on terms.'

But the body is a good example of the difficulty of being quite sure about tensegrity. There are many structures in which some parts are in tension and others in compression – the Forth Bridge for example – but they are not tensegrity structures because the compression members are all in contact. The skeleton is certainly held together by the pre-stressed tension members and it is true that the bones don't touch each other, but besides floating in a springy net of tension members, the bones butt up against a cartilage pad which lies between them.

Vorticella, the tiny self-deploying organism that can either squash itself up into a ball or extend itself 2–3 mm on a slender thread, may be able to shed further light on tensegrity's biological role. When it is extended, a network of stiff molecular filaments is maintained in a rigid form by electrical repulsion. The filaments are held taut as in a

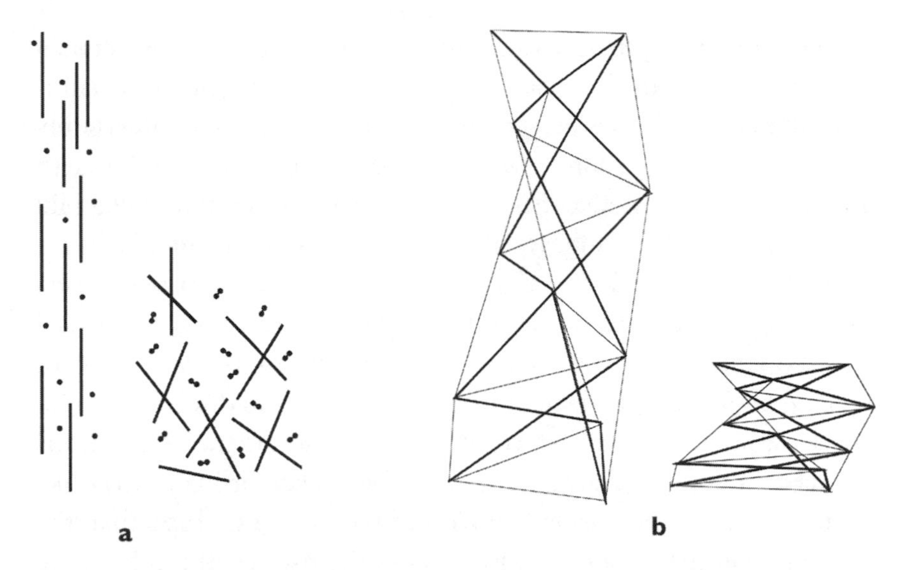

Fig. 9.12 Deployable tensegrity structures in nature and technology. When extended, the tiny aquatic organism *Vorticella* (a) maintains its shape through electrical repulsion; it can collapse itself by neutralizing the repulsion. Sergio Pellegrino's tensegrity mast (b) deploys and retracts in a similar manner.

---

tensegrity sculpture and they do not touch because they are mutually repelling. So it is this tension force-field that holds the filaments in place. It is sometimes said that a true tensegrity structure would not be useful either in architecture or biology because if the continuous tension wire is broken at any point, the whole network collapses. *Vorticella*, it seems, has made a virtue of this and in this case the process is fully reversible.

There are large-scale engineered tensegrity structures that use the kind of deployable mechanisms seen in Chapter 8. Sergio Pellegrino at the University of Cambridge has designed masts for space arrays using erectable tensegrity structures. A schematic diagram of the *Vorticella* filament structure looks very like a deployable tensegrity structure (fig. 9.12).

Kenneth Snelson believes that by now the waters have been so muddied that no one will ever really know now what tensegrity is. But between Fuller's 'everything is tensegrity' and Snelson's 'only my sculptures are tensegrity' there is much interesting subject matter for debate. The principle of discontinuous compression is intellectually and aesthetically thrilling, as is its embodiment in Snelson's sculpture; the Geiger domes are practical, beautiful, cheap and certainly closer to tensegrity than to any other principle. As I left Kenneth Snelson, I suggested that perhaps at least the human skeleton did exhibit the property of tensegrity? 'Almost' was his laconic reply.

A few days after I met Kenneth Snelson, I went to see Moshe Safdie's Peabody Essex museum in Salem, Massachusetts. Safdie is a Canadian architect who explicitly acknowledges the influence of D'Arcy Thompson, and the museum dispelled the doubts Kenneth Snelson's scepticism had sown in my mind about the connections between biology and architecture. Safdie makes brilliant atria-like roofs and here the forms echo both the whaling ships that the museum commemorates and the whales themselves. It is a building for the age of bio-inspiration.

It is understandable that the relationship between biology and architecture should be hard to pin down. But architecture is as much a symbolic art as a practical one. Since nanotechnology, for all its marvels, is largely hidden, it is good to have some large structures that echo the principles of natural design and make them flagrantly visible to all. This is not retro-architecture: it has classical grace but it uses clean lines that derive from the nature of the materials and it acknowledges that modernism happened. Architecture should be of its time and this style proclaims that we are a vital civilization in our own right and not just the people who, thanks to technology, are able to archive everything that had happened before us.

This architecture is highly congruent with the new bio-inspired materials and will increasingly come to incorporate them – Activ glass being the first example. Although Julian Vincent's argument that the organicism of recent architecture is largely skin-deep has some force, there is no doubting the parallels between structures such as geodesic domes, Isler's shells and natural structures. David Campbell says: 'We're trying to get less out of more, we're trying to get multiple functions and a level of integration.' Maximum economy, multiple functions and integration have been precisely nature's strategies throughout evolution.

CHAPTER TEN

Designing the Future (Naturally)

> What new Americas of light have been
> Yet undiscovered there, or yet unseen,
> Art's near approaches awfully forbid,
> As in the majesty of nature hid.
>
> RICHARD LEIGH, 'Greatness in Little'

Bio-inspiration represents an attitude to life that will be expressed in many different ways beyond the merely practical. It suggests a light aesthetic – in every sense, not just visual. You can find this sense of lightness in Lucretius, who has been a constant touchstone throughout the book, and the spirit of Lucretius was memorably evoked by Italo Calvino in *Six Memos for the Next Millennium*:

> The *De Rerum Natura* of Lucretius is the first great work of poetry in which knowledge of the world tends to dissolve the solidity of the world, leading to a perception of all that is infinitely minute, light and mobile.

'Infinitely minute, light and mobile' would describe many of the subjects of this book. The larger structures described in the preceding two chapters cannot be infinitely minute but they are certainly light and some of them are mobile. Heaviness, both literally and spiritually, is the enemy. It is early days for bio-inspiration but there are signs that aesthetically it will be welcomed. For there is a turning away from the old, the heavy, the steeped-in-history. The new signature buildings of architects such as Santiago Calatrava and Frank Gehry are popular, as the antiques trade begins to struggle. In the search for this new lighter aesthetic, people are finally casting off the last remnants of Victorian clutter – the love of Gothic gloom and bric-a-brac. People are discovering the joy of living in their own time.

Nature and technology have been seen as polar opposites by many, with nature as a comforting balm for gritty hard-edged city technology. But nature at the nanolevel looks like ... well, contemporary high technology and architecture! So the nature/technology antithesis breaks down in the face of the new science and technology. This is surely a good omen for the future. It signals a culture less divided, less neurotic about the natural and the synthetic, less timid and backward looking.

Bio-inspiration can contribute to a new aesthetic for our times in two ways. Firstly, in uncovering the mechanisms of some of nature's most impressive and beautiful phenomena, it has greatly enlarged the scope of biology. The old, unhelpful divisions between molecular biology and the natural-history tradition, including fieldwork and the classification of species, is breaking down. Biology is becoming a complete and unified science. And to reiterate the essential thesis of Richard Dawkins's book *Unweaving the Rainbow*: to understand the mechanisms of nature does not detract from the poetry. In bio-inspiration, not only does the science not detract from the poetry of nature, the science has *uncovered* hidden poetry. For millions of years the optical structures in the sea mouse and *Morpho* butterflies bent light in ways unknown to humanity. Bio-inspiration has opened up a new realm of nature as surely as did the coming of the microscope or the unravelling of the structure of DNA in 1953.

The second contribution lies in the potential applications of this work. Now we are on more difficult territory. Hubris is a common human weakness. No sooner have we *begun*, tentatively, to understand some natural mechanism than claims are made for instant technical nirvana. Almost every article written about bio-inspiration in the last few years has ended with a sentence to the effect that: 'Soon we shall all be doing it the gecko's/spider's/butterfly's way.' I confess that some of them had my name on the byline. In the modern journalistic climate, it seems to be necessary to make this kind of pitch to win attention. And hype also unlocks research funding. As the work in this book has shown, the reality has been very different, with many setbacks and slower progress than anticipated.

This is no different to any other cutting-edge science. In the early 1980s, when genetic engineering became a reality, we were promised gene therapy within a few years. Some disastrous attempts *were* made within that time but, 20 years on, it all appears to be vastly more complicated than it seemed at first. During the timespan of writing this book, the fate of the individual projects waxed and waned unpredictably. Spider silk, which was trumpeted in the mainstream press as a success story about to deliver, received heavy setbacks, whereas photonic crystals, the revolutionary way of guiding light for computer optics, invented in 1987 and discovered to exist in many natural creatures, began to glow with promise. This process will go on. George de Mestral had to wait a decade or so for commercial nylon to enable him to produce his hook-and-loop fasteners but technologies can suffer even worse delays: the computer, effectively invented by Babbage in the 1830s, was still-born for over a century for lack of the science of electronics. Some of bio-inspiration's ideas will similarly have to await their enabling technology. Having said all that, the range of techniques available now is so rich that a researcher stymied in one direction has plenty more to call on. But what we have yet to see emerge are standard generic techniques such as those used in silicon-chip manufacture.

Of the two kinds of bio-inspirationists, the materials scientists tend to be bullish because their science is a new frontier, just as molecular biology was in the 1960s. Biomechanics is an older discipline and its practitioners are more wary. As usual, I would guess that the true position lies somewhere between these poles. Applications there will certainly be but, except in a few cases, the gestation will be long – typically up to 20 years – and fraught with setbacks. That we have learnt so much about this nanorealm of nature does not mean that we shall soon know it all. D'Arcy Thompson, who made nature's engineering his life's work, at a time when clear answers were very hard to come by, said:

> That nature keeps some of her secrets longer than others – that she tells the secret of the rainbow and hides that of the northern lights – is a lesson taught me when I was a boy.

There is a pattern in several of the stories, of an initial breakthrough leading to premature optimism. Nature's mechanisms are subtle and complex but sometimes a crude approximation gives surprisingly good results – then the going starts to get tough. Spider silk and mussel glue are two such cases in point. Professor Bob Full insists that a slavish copying of nature will fail: you can only start to do something similar to nature's work when you understand the underlying principle. And then what you make might be very different to the natural version. In the case of spider silk, you could argue that we had already abstracted the principle of spider silk when we made nylon – we can get by without the extra properties conjured by the spider. Nevertheless, getting to the bottom of these mechanisms must be as valid a piece of research as investigating black holes or pursuing the superstring theory of matter.

Much research will never lead to any technical product, but it is surprising how much of what once seemed innocuous, arcane and – to many people – rather pointless biological research does now have the prospect of potential commercial development. Early 20th-century papers on the gecko's foot or structural colour in butterflies did not look as if they would ever be of interest to technologists.

Human technology has a long history and a very broad range of techniques: bio-inspiration is about 15 years old, although it has deeper roots in Germany, perhaps going back to the 1960s. Not the least of its virtues is that it gives an enormous boost to arguments in favour of conservation and biodiversity. The argument for biodiversity – apart from maintaining the general health of the planet and providing for human recreation – has always focused on the potential drugs that could exist in as yet unknown species and on the need to preserve the gene bank of economic crops.

But now, creatures that were formerly thought to be merely cute or weird, and to be preserved just for their oddity, turn out to be blueprints for entire new technologies. Geckos, spiders and flies formerly had no use at all or were regarded as pests. If geckos had been rendered extinct before the era of the electron microscope, it is unlikely that the adhesive mechanism they employ would ever have been discovered. And there could be species out there, harbouring nanosecrets, that are disappearing as I write. The more pointless bits of creation (from our point of view) just got more pointed.

Scientists don't always speak with one voice, especially in an emerging multi-disciplinary subject. Time and again, talking to the scientists, disagreement was not far beneath the surface. In many cases, the right way to do these things has not yet emerged and the scientists are all too human in their need to defend their own patch and cast a surly eye on others who do things differently. Without putting names to phrases, 'That's nonsense!' 'That's not data!' 'But that would be *chemistry*!' 'He doesn't know anything about it!' were a few of the noises off. Those who want science to be cut and dried will probably be disgusted by this but cantankerousness is all part of the passionate search for truth. And once definitive results are obtained, all will accept the result and some will admit they were wrong.

In this book, I have written about science and technology on the hoof because this is a new science and success is a matter of decades of work. Rather than wait until some of these technologies have become commonplace, I have tried to capture the Wordsworthian 'bliss-it-was-in-that-dawn-to-be-alive' moment of seeing what was, until 15 years ago, a wholly unexpected science take shape before our eyes.

Across the wide range of techniques covered in this book there is one overriding message: shape, shape, shape. At every size, from atoms to cabledomes, we see that shapes can do things we could hardly have expected. There are some who think that the three classical dimensions are boring: mathematicians play with geometries of any number of dimensions and at the further reaches of physics three dimensions no longer cut it.

But whether it is in surprising ways to get from folded to flat (Miura-ori), how to be self-cleaning (Lotus-Effect), how to stick without being sticky (the gecko's foot), how to bend light to our purpose (photonic crystals), how to self-assemble electronic components (the molecular erector), how to raise a roof with tensile wires that ought to sag rather than soar (tensegrity) – in all of these it is the patterns these structures make in space rather than what they are made from that creates the effect. Galileo's intuition, from the birth of modern science in the 17th century, that the book of nature is written in circles and squares and triangles, has been vindicated, although some of the shapes are far more complicated than he could ever have foreseen.

NOTES

Chapter 1

p. 3 'There's plenty of room at the bottom': *Engineering and Science*, 'There's plenty of room at the bottom', February 1960, pp. 22–36. The lecture is available in Feynman (2001), pp. 117–39, and on the web at http://www.zyvex.com/nanotech/feynman.html

p. 8 'What we can *see* with the aid of a microscope': Ramsay (1965), pp. 55–62.

p. 8 'When light hits objects patterned at just below one thousandth of a millimetre': Jones (2004), pp. 78–80. Jones's book, principally about nanotechnology but with a bias towards biology, is a readable account of many aspects of the subject.

p. 9 'Brief reflection on cats growing in trees': Holub (1990), p. 144. Miroslav Holub (1923–98) was that rare beast, a distinguished practitioner of both science (immunology) and poetry. Writing against the backdrop of communist Czechoslovakia, he was one of the most valuable witnesses to late 20th-century civilization.

p. 12 'The pictures revealed by the SEM': Claugher (1990).

p. 12 'A kind of delicate meeting place between imagination and knowledge': Nabokov (2000). Nabokov's insight is shared by many poets and writers. Bachelard (1969) discusses the idea of 'intimate immensity' in Baudelaire's poetry.

p. 13 'The Spirit is too Blunt an Instrument': Stevenson (2000), p. 24.

p. 13 'Some estimates go as high as 30 million or more': *Philosophical Transactions of the Royal Society B*, 29 November 1990, pp. 171–82. This is scientific data at its most imprecise – estimates of actual named species are approximately 1.8 million. Obviously, no one expects to find thousands of unknown large creatures but the smaller things get the more niches for species there seem to be.

p. 13 'The more than 24 million known chemical combinations': there is no real limit to the number of chemical combinations, especially those involving the carbon atom (carbon chemistry is known as organic chemistry, irrespective of whether the compounds concerned occur in living things). Not every possible combination would have useful properties, either in living things or technology.

Chemical Abstracts Service (CAS), a division of the American Chemical Society, is the standard world source for chemical documentation. Every new chemical compound or biological molecule sequence is allocated a CAS Registry Number. At the time of writing, besides the 24 million compounds, there were almost 50 million sequences of biological molecules. These figures grow very rapidly and updates can be seen at www.cas.org/cgi-bin/regreport.pl

p. 14 'Small things must be simple': *Applied Optics*, 'Light and Color on the Wing', 20 August 1991, p. 3499.

p. 14 'Everything around us was a mystery': Levi (2000), p. 19. Like Miroslav Holub, Primo Levi (1919–87) was a working scientist, writer and a man with a distinctive perspective on the tragic history of the 20th century. Levi was a chemist who as an Italian Jew was deported to Auschwitz. He survived and became a celebrated writer, both for his accounts of Auschwitz and for his 'curious roaming in other people's trades'.

p. 15 'Nylon for instance, invented in 1937': van Dulken (2000), pp. 62–3.

p. 15 'Natural silks have different amino acid units, linked nose to tail': for basic silk protein science *see* Asquith (1977); for a more recent and highly technical survey of silk protein structures, *see* Kaplan et al. (1994).

p. 15 'All science is either physics or stamp collecting': Birks (1962), p. 108.

p. 15 'One gram of carbon contains about fifty thousand million million million atoms': this value was first calculated by Johann

Josef Loschmidt who, in 1865, used the 18th-century chemist Amadeo Avogadro's idea that equal volumes of gases always contained the same number of molecules to calculate the number of atoms in a set quantity of any element. Jean Baptiste Perrin named this quantity Avogadro's number in 1908. The set quantity used is the mole which is the gram equivalent of the atomic or molecular weight – for instance, for carbon it is 12, for oxygen 32 (oxygen existing as the molecule O_2).

Avogadro's number is very large – 6.022×10^{23}, so it follows that the atoms must be very small. (http://scidiv.bcc.ctc.edu/wv/6/0006-003-avogadro.htm) From Avogadro's number we can calculate the size of an atom. If we measure the density of the substance, the volume of the atom is given by the equation:

Atomic volume – molar mass in grams/density x Avogadro's number

The result is that the carbon atom has a diameter of about 0.3 nanometre. (http://hyperphysics.phy-astr.gsu.edu/hbase/particles/atomsiz.html)

p. 17 'It is not a living thing': Hardy (1956), p.94.

p. 19 'The most dramatic recent example has been the photonic crystal': for a survey of this burgeoning new field, *see* Joannopoulos et al. (1995).

p. 20 'These small flying fortresses': Levi (1989), pp. 17–18.

p. 20 'The flashing light of the firefly': Harvey (1952); Nealson (1981).

p. 20 'A desert beetle': *Nature*, 'Water capture by a desert beetle', 1 November 2001, pp. 33–4.

p. 21 'The bombardier beetle': *Science*, 'Defensive spray of the bombardier beetle: a biological pulse jet', 8 June 1990, pp. 1219–21. So technical is the bombardier's mechanism it has been seized upon by creationists in the USA as an example of 'something that could not possibly have evolved'.

p. 21 *De Rerum Natura*: Lucretius (1969).

p. 22 'Under the SEM a natural silk fibre will be observed to have micro-structured rough edges': Hongu and Phillips (1997), p. 44.

p. 22 'Greatness in Little': Leigh (1947), pp. 22–4.

p. 23 'Hybrid technologies': *Nature Materials*, 'Molecular biomimetics: nanotechnology through biology', September 2003, pp. 577–85.

p. 23 'A coded blueprint strung out along the double helix of DNA': Jones (2004), pp. 110–13.

p. 24 *Prey*: Crichton (2002).

p. 24 *Engines of Creation*: Drexler (1986).

p. 25 'Prince Charles in a speech warning of the dangers of nanotechnology': *Mail on Sunday*, 27 April 2003.

p. 25 'Report on the benefits and possible dangers of nanotechnology': The Royal Society and the Royal Academy of Engineering, 'Nanoscience and nanotechnologies: opportunities and uncertainties', 29 July 2004.

p. 27 'Expo 2005': 'Nature's Wisdom', held at Aichi, Japan, from 25 March to 25 September, 2005. http://www.expo2005.com/

p. 27 'The *Zoomorphic* exhibition at the Victoria and Albert Museum': the exhibition lives on in the form of a book, written by the exhibition's creator Hugh Aldersey-Williams; Aldersey-Williams (2003).

Chapter 2

p. 29 'To the Lotus-Bloom': Komai (1927), pp. 157–8.

p. 30 'The Threefold Lotus Sutra': Kato (1975), p. 14.

p. 31 'No more stately plant adorns our gardens than lotuses': Conard and Hus (1907), p. 157.

p. 31 *fn* 'In evolutionary terms, lotuses and water lilies are not related': Barthlott (1998), pp. 408–16.

p. 33 'The surface of plants is a strange other-worldly terrain': Barthlott (1990), pp. 69–94.

p. 34 'Many plants never seemed to need cleaning': *The Sciences*, 'The Lotus Effect', January/February 2000, pp. 12–15.

p. 34 *fn* 'The strangest thing about the waxes': *Planta*, 'Movement and regeneration of epicuticular waxes through plant cuticles', July 2001, pp. 427–34.

p. 35 'The self-cleaning effect depends on the relative "wettability"': Barthlott (1990), p. 87.

p. 36 'The Fakir-on-the-Bed-of Nails': a physicist's look at the Fakir Effect can be found at http://www.fas.harvard.edu/#scdiroff/lds/

NewtonianMechanics/FakiPhysics/FakirPhysics.html

p. 36 'The self-cleaning effect was most noticeable in the sacred lotus': *Planta*, 'Purity of the sacred lotus or escape from contamination in biological surfaces', vol. 202, no. 1, 1997, pp. 1–8.

p. 38 'In 1992, Barthlott established the name Lotus-Effect®': *Klima- und Umweltforschung an der Universität Bonn,* Rheinische Friedrich-Wilhelms-Universität Bonn, 'Die Selbstreinigungsfähigkeit pflanzlicher Oberflächen durch Epicuticularwachse', 1992, pp. 117–20.

p. 39 'The Lotus-Effect officially entered the canon of Western inventions': International Patent WO 96/04123, granted 1998.

p. 39 'The classic summing up of the Lotus-Effect': *Planta*, 'Purity of the sacred lotus or escape from contamination in biological surfaces', vol. 202, no. 1, 1997, pp. 1–8.

p. 42 'Spray-on temporary Lotus-Effect formulations': International Patent WO 00/58410.

p. 44 'Titanium dioxide also has unusual electro-optical properties': Fujishima et al. (1999).

p. 47 '*Which?* magazine gave Activ glass a brief write-up': *Which?*, 'Self-cleaning Windows', June 2003, p. 53.

p. 48 'Lotus-Effect coatings from poly-propylene', *Science*, 'Transformation of a Simple Plastic into a Superhydrophobic Surface', 28 February 2003, pp. 1377–80.

p. 48 'A self-cleaning fabric known as Nano-Care®': US Patent 2003/0,013,369, 16 January 2003.

p. 50 'Could titanium dioxide be used with Lotus-Effect coatings to produce a self-renewing capability?': *Langmuir*, 'Self-cleaning hydrophobic surfaces', 22 August 2000, pp. 7044–7.

p. 50 'Nature has already combined the Lotus-Effect and Activ technology': *Nature*, 'Water capture by a desert beetle', 1 November 2001, pp. 33–4.

p. 51 'An efficient new way of collecting water': US Patent 2004/0,109,981, 10 June 2004.

p. 52 'Non-stick water': *Nature*, 'Liquid marbles', 21 June 2001, pp. 924–7.

p. 53 'The aphid's secret': *Proceedings of the Royal Society of London B*, 'How aphids lose their marbles', 22 June 2002, pp. 1211–15.

p. 54 '*The Poetics of Space*': Bachelard (1969), p. 67.

Chapter 3

p. 55 'The astounding properties of spider silk': Foelix (1996), pp. 110–12.

p. 56 'The oldest existing strand of spider silk': *Nature*, 'Palaeontology: Spider-web silk from the Early Cretaceous', 7 August 2003, pp. 636–7.

p. 57 'Tropical spider webs can be very large': Pratt (1906), pp. 266–7.

p. 57 'In 1725, Sir Hans Sloane reported the nets were "so strong as to give a man inveigled in them trouble for some time"': Hillyard (1994), p. 100. Hillyard's book is the best popular natural history of all aspects of spiders in life, science and art.

p. 57 'The old cobwebs in cellars and attics': Levi (1989), p. 144.

p. 58 'The first documented attempt to exploit spider silk was by a Frenchman': Bon (1748). Bon wrote, 'One is surprised to learn that spiders make a silk as fine, as strong and as lustrous as common silk; the prejudice against an insect as well known as it is despised is due to the fact that people have not known until now the use that can be made of it.'

p. 58 'René Réaumur investigated these claims': Hillyard (1994), p. 121.

p. 60 'A garden spider has three pairs of spinnerets': Foelix (1996), pp. 117–22.

p. 60 'Probably there might be a way found out, to make an artificial glutinous composition': Hooke (1665), p. 7.

p. 60 'For centuries the only way of making silk was with the silkworm': Asquith (1977), pp. 53–80.

p. 61 'A parasitic disease called *pébrine*': Baricco (1997), p. 7.

p. 62 'The potential of silks in one of the toughest applications imaginable': *Southern California Practitioner*, 'Notes on the Impenetrability of Silk to Bullets', vol. 11, 1887, pp. 95–8.

p. 62 'The invention of nylon': van Dulken (2000), pp. 92–3. Nylon almost did not make it as one of the great inventions of the 20th century. Wallace Hume Carothers first produced it in 1933 but it could not be spun into useful fibres. Four years later, while working with another new fibre – the first polyesters – Carothers' co-workers discovered that tough threads could be made by drawing out threads from the raw sticky polyester. They went back to try the same with nylon and the rest is history. But the success came too late for Carothers. Three weeks after his nylon patent (US 2,130,948) was filed he committed suicide.

p. 62 'Kevlar, a tougher variant of nylon, invented in 1963': van Dulken (2000), pp. 154–5.

p. 63 'To produce industrial quantities of spider silk from the milk of genetically engineered goats': *Science*, 'Spider silk fibers spun from soluble recombinant silk produced in mammalian cells', 18 January 2002, pp. 472–6; *Guardian*, 'Spinning goats and a web of intrigue', 18 January, 2002.

p. 63 'We use water and hay': *Style.ca*, September 2003; http://www.style.ca/PreviousIssues/Sept2003/UpClose.jsp

p. 64 'Spider silk is a protein': for a good introduction to proteins and DNA, *see* Niklos and Freyer (2003), pp. 48–58. The second half of this book is for the experimenter but the first half is a fine primer in DNA and protein science.

p. 66 'Domains of strongly orientated molecules that act as liquid crystal zones': *Nature*, 'Liquid crystalline spinning of spider silk', 29 March 2001, pp. 541–8.

p. 67 'The most plausible model for the structure of spider silk': *Scientific American*, 'Spider Webs and Silks', March 1992, pp. 52–5. Although seemingly out of date, this paper by Fritz Vollrath, one of the great experts on spider silk, is still one of the clearest popular accounts.

p. 68 'There is a whole science of cracks': Gordon (1991), pp. 65–9.

p. 70 'Randy Lewis's laboratory at the University of Wyoming': *Trends in Biotechnology*, 'Synthetic spider silk: a modular fibre', September 2000, pp. 374–9.

p. 72 'Spinning silk from "amphiphilic polymers"': *Nature*, 'Liquid crystalline spinning of spider silk', 29 March 2001, pp. 54–8; *Philosophical Transactions of the Royal Society B*, 'Biological liquid crystal elastomers', 28 February 2002, pp. 155–63. This is a themed issue on elastomeric proteins, including collagen and elastin as well as spider silk.

p. 72 'A spinning nozzle that can handle many different solutions':

International Patent WO 01/38614 (2001).

p. 73 'A process for making various protein polymers, such as keratin and wool, soluble': US Patent 6,034,220 (2000).

p. 74 'A neat way of demonstrating the effect of moisture on spider silk': *Materials Today*, 'Steamed superhero?', July/August 2004.

p. 75 'An important paper on spider-silk protein sequences': *Journal of Molecular Biology*, 'Mapping Domain Structures in Silks from Insects and Spiders related to Protein Assembly', 2 January 2004, pp. 27–40.

p. 75 'Silkworm silk is not as tough as spider silk': this is true of the natural state but Fritz Vollrath at Oxford University showed that by varying the speed of reeling from silkworms silks almost as strong as spider silk can be produced. This is important evidence for the importance of the physical spinning process in the making of strong fibres from natural proteins: *Nature*, 'Surprising strength of silkworm silk', 15 August 2002, p. 741.

p. 76 'Control silk spinning through the water content alone': *Nature*, 'Mechanism of silk processing in insects and spiders', 28 August 2003, pp. 1057–61.

p. 76 'A Small Business Technology Transfer Program grant': US Department of Defense Small Business Technology Transfer Program Phase 1, 2001, ID#F013 – 0111, topic 010014.

p. 77 'A follow-up grant': US Army Research Office, 2003, A03-054.

p. 78 'The fine structure of spider's feet': *Proceedings of the National Academy of Sciences of the USA*, 'From micro to nano contacts in biological attachment devices', 16 September 2003, pp. 10603–6.

Chapter 4

p. 79 'Geckos have always astonished everyone who has ever seen them – and that includes Aristotle': Aristotle makes a passing reference to the gecko in *Historia Animalium*; the text, translated by the pioneer of bio-inspiration D'Arcy Thompson, can be found at http://classics.mit.edu/Aristotle/history_anim.html

p. 79 'They are a group of nocturnal lizards': Pianka and Vitt (2003), pp. 171–85. Apart from pet manuals, geckos are not well covered in the literature but this is an attractive, up-to-date (it mentions Autumn's work on gecko adhesion) full-colour work.

p. 83 'Autumn's team published a paper in *Nature*': *Nature*, 'Adhesive force of a single gecko foot-hair', 8 June 2000, pp. 681–4.

p. 84 'Van der Waal's force': a good explanation of van der Waals' forces can be found at www.chemguide.co.uk/atoms/bonding/vdw.html.

p. 86 'Some pretty good synthetic gecko arrays': *Journal of Adhesion Science and Technology*, 'Synthetic Gecko Foot-Hair Micro/Nano-Structures for Future Wall-Climbing Robots', vol. 17, no. 8, 2003, pp. 1055–73.

p. 86 'Different results would be expected on strongly water-repelling and water-attracting surfaces': *Proceedings of the National Academy of Sciences of the USA*, 'Evidence for van der Waals adhesion in gecko setae', 27 August 2002, pp. 12252–6.

p. 87 'Gecko tape': *Nature Materials*, 'Microfabricated adhesive mimicking gecko foot-hair', July 2003, pp. 461–3.

p. 89 'The gecko mechanism was patented': US Patent 6,737,160, granted 18 May 2004.

p. 91 'The *larger* the creature the *finer the division of the bristles*': *Proceedings of the National Academy of Sciences of the USA*, 'From micro to nano contacts in biological attachment devices', 16 September 2003, pp. 10603–6.

p. 92 'Velcro® brand hook-and-loop fastener': van Dulken (2000), pp. 144–5.

p. 93 'A probabilistic fastening': Nachtigall (1974), pp. 68–70. Nachtigall's book is an encyclopaedic survey of biological plugs and sockets, suction cups, penile expansion nuts and the like.

p. 94 'De Mestral's grandson': *La Côte*, 9 July 2004.

p. 95 'The original patent': GB 721,338, filed 1951.

p. 96 'Difficulties with the Californian Velcro crop': http://home.inreach.com/kumbach/velcro.html

p. 97 'The mussel's secrets are being prised out': *Macromolecules*, 'Synthetic Polypeptide Mimics of Marine Adhesives', 28 July 1998, pp. 4739–45; *Angewandte Chemie International Edition*, 'Metal-mediated Cross-Linking in the Generation of a Marine-Mussel Adhesive', 4 January 2004, pp. 448–50.

p. 98 'The adhesive powers of DOPA have been put to a kind of reverse use': *Journal of the American Chemical Society*, 'Mussel

Adhesive Protein Mimetic Polymers for the Preparation of Nonfouling Surfaces', 9 April 2003, pp. 4253–8.

Chapter 5

p. 101 'The eye is thought to have evolved on 40 separate occasions': Dawkins (1996), p. 127.

p. 101 'Light became an evolutionary force over 500 million years ago': Parker (2003).

p. 102 'The Pentagon has commissioned a study': *Guardian*, 'Life', 'Natural Defences', 18 November 2004, p. 6.

p. 102 'An all-optical computer': *Journal of Materials Chemistry*, 'Towards the synthetic all-optical computer: science fiction or science reality?', 7 March 2004, pp. 781–94.

p. 102 'The photonic crystal': *Physical Review Letters*, 'Inhibited Spontaneous Emission in Solid-State Physics and Electronics', 18 May 1987, pp. 2059–62; *Physical Review Letters*, 'Strong Localization of Photons in Certain Disordered Dielectric Superlattices', 8 June 1987, pp. 2486–9.

p. 104 'A large-scale hollow crystal': *Physical Review Letters*, 'Photonic band gap structure: The face-centered-cubic case employing nonspherical atoms', 21 October 1991, pp. 2295–301.

p. 104 'Many kinds of butterflies and some marine creatures, even the occasional beetle, were going about their business': *Nature*, 'Photonic structures in biology', 14 August 2003, pp. 852–5; *Journal of Optics, A: Pure and Applied Optics*, '515 million years of structural colour', November 2000, R15–28.

p. 105 'The finely colour'd Feathers of some Birds': Newton (1931), p. 252.

p. 106 'There were many times of an evening': Mann (1999), p. 17.

p. 110 'Electron microscope pictures and drawings that revealed a fantastic Piranesian world of galleries and cells': *Applied Optics*, 'Light and color on the wing', 20 August 1991, pp. 3492–500.

p. 110 'The widespread distribution of iridescence': *Applied Optics*, 'Light and color on the wing', 20 August 1991, p. 3497.

p. 110 'The bright blue *Morphos*': D'Abrera (1984), pp. 331–73. This brilliantly illustrated book shows the *Morphos* in their full glory. It also discusses the confusion surrounding the species of this genus.

The author says: 'I have tried to get away from the "stamp-collecting" syndrome that has developed around this group.'

p. 111 'The *Morpho* wing is covered in scales': *Proceedings of the Royal Society of London B*, 'Quantified interference and diffraction in single *Morpho* butterfly scales', 22 July 1999, pp. 1403–11.

p. 112 'The iridescence of something that rejoices both in the hum-drum common name of the sea mouse': *Nature*, 'Aphrodite's iridescence', 4 January 2001, pp. 36–7.

p. 113 'The photonic crystal fibres invented in 1995 by Philip Russell': *Science*, 'Photonic Crystal Fibres', 17 January 2003, pp. 358–62. A recent authoritative review of the subject by the inventor of two-dimensional photonic crystal fibres.

p. 114 'Another sea worm – *Pherusa*': Vukusic (2003), p. 24.

p. 115 'The South American butterfly *Parides sesostris*': Vukusic (2003), pp. 24–6.

p. 118 'Inverse opal photonic crystals': *Advanced Functional Materials*, 'Opal Circuits of Light – Planarized Microphotonic Crystal Chips', June 2002, pp. 425–31.

p. 118 'Anti-counterfeiting devices': *Physics World*, 'Diffraction, beauty and commerce', October 1989, pp. 24–8. An article well ahead of its time in predicting the impact of butterfly optics on optical security devices. A wide-ranging overview of this fascinating field, with some stunning iridescent visuals, can be found in van Renesse (1998).

p. 119 'Polarized light was shown to be used for finding mates': *Nature*, 'Polarized light as a butterfly mating signal', 1 May 2003, p. 31.

p. 119 '*Ancyluris meliboeus* creates a flickering effect': *Proceedings of the Royal Society of London B*, 'Limited-view iridescence in the butterfly *Ancylurus meliboeus*', 7 January 2002, pp. 7–14.

p. 119 'The extreme dependence of this iridescence on angle makes it an attractive technical proposition': US Patent Application 2004/0,101,638.

p. 119 'The most famous bio-inspired reflector is the Catseye®': van Dulken (2000), pp. 86–7.

p. 121 'Butterflies and moths have a finely structured array of tiny bumps across the cornea': *American Scientist*, 'Light Reflection

Strategies', May–June 1999, pp. 252–5.

p. 122 'F117 Stealth bomber': *American Scientist*, 'Light Reflection Strategies', May–June 1999, p. 252.

p. 123 'The mysteries of mimicry had a special attraction for me': Nabokov (2000), p. 98.

p. 123 'The orange and green cicadas': von Frisch (1973), pp. 87–8. In some species, some members have green wings and others have yellow/orange/ red wings. They perch *en masse* on twigs, creating the illusion of flower spikes.

p. 124 'The resemblance is literally superficial': Cott (1957), p. 324. Cott's book, originally published in 1940, is remarkably bio-inspired in tone, and a dazzling compendium of the apparently 'artistic' disguises adopted by many species.

p. 125 'In Tachi's experiment a person wears a hooded cloak': *Observer*, 'Japanese boffins spawn almost invisible man', 13 June 2004. Details of Tachi's experiment can be found at: http://projects.star.t.u-tokyo.ac.jp/projects/MEDIA/xv/oc.html

p. 126 'Creatures that advertise the fact that they are dangerous': von Frisch (1973), pp. 91–2.

p. 126 'The brittlestar story began with Gordon Hendler's fieldwork': *Zoomorphology*, 'Fine structure of the dorsal arm plate of *Ophiocoma wendtii*: Evidence for a photoreceptor system', vol. 107, 1987, pp. 261–72.

p. 129 'The base of the tower is encircled with fine spines': *Nature*, 'Fibre-optical features of a glass sponge', 21 August 2003, pp. 899–900.

p. 130 'Aizenberg's work on the brittlestar': *Nature*, 'Calcitic micro-lenses as part of the photoreceptor in brittlestars', 23 August 2001, pp. 819–22.

p. 131 *fn* 'Calcite lenses similar to those of the brittlestar have also been found in fossil trilobites': *Science*, 'Trilobite eyes: Calcified lenses in vivo', 9 March 1973, pp. 1007–10.

p. 132 'There have been various studies of the peacock over the years': *Proceedings of the National Academy of Sciences of the USA*, 'Coloration strategies in peacock feathers', 28 October 2003, pp. 12576–8.

Chapter 6

p. 136 'This process is called "biomineralization"': Mann (2001).

p. 136 *fn* 'The affinity between proteins and clays goes very deep': Cairns-Smith (1990).

p. 137 'The intricate patterns on a silicon chip': *Scientific American*, 'The first nanochips', April 2004, pp. 48–55.

p. 137 'Computer chips would double in power about every 18 months': *Electronics*, 'Cramming more components onto integrated circuits', 19 April 1965, pp. 114–16. There is some dispute as to how well Moore's Law has been followed in practice. Many reports claim a doubling every 18 months, although Moore himself has repudiated this idea. For a discussion of this question, see *First Monday*, 'The Lives and Death of Moore's Law', November 2002, available online at: http://firstmonday.org/issues/issue7_11/tuomi/index.html

p. 137 'The computer industry has an industry-wide standard roadmap': International Technology Roadmap for Semi-conductors. Available online at: http://www.itrs.net/Common/2004Update/2004Update.htm

p. 138 'How DNA makes structures': Jones (2004), pp. 110–12.

p. 138 'Proteins *self-assemble*': Jones (2004), pp. 113–20.

p. 140 'Abalone nacre reveals a composite structure': Baeurlein (2000), pp. 221–49.

p. 141 'Abalone's mineral synthesis could be induced artificially': *Nature*, 'Flat pearls from biofabrication of organized composites on inorganic substrates', 1 September 1994, pp. 49–51.

p. 143 'New proteins that can recognize and preferentially stick to computer-chip materials': *Nature Materials*, September 2003, pp. 577–85. A review article on protein templating and phage display techniques by Mehmet Sarikaya and colleagues.

p. 143 'Phages and their amazing properties': *Scientific American*, 'Building a Bacterial Virus', July 1967, pp. 61–74.

p. 143 'But then, most dreadful of all, the phage': Holbrook (1980), p. 138.

p. 146 'Electrical components with solder connections can wire themselves up': *Science*, 'Forming Electrical Networks in Three Dimensions by Self-Assembly, 18 August 2000, pp. 1170–2.

p. 146 'An array of 1,500 silicon cubes': *Science*, 'Fabrication of a Cylindrical Display by Patterned Assembly', 12 April 2002, pp. 323–5.

p. 147 'Angela Belcher discovered a brilliant shortcut': *Science*, 'Ordering Quantum Dots Using Genetically Engineered Viruses': 3 May 2002, pp. 892–5.

p. 148 'Self-assembling properties can be used to create structures without the phage': *Nature Materials*, 'Molecular biomimetics: nanotechnology through biology', September 2003, pp. 577–85.

p. 150 'The ATP motor': *Science*, 'Powering an Inorganic Nanodevice with a Biomolecular Motor', 24 November 2000, pp.155–8.

p. 152 'An elegant new way of harnessing biological energy': *Nature Materials*, 'Self-assembled microdevices driven by muscle', February 2005, pp. 180–4.

p. 153 'Ways of modulating the crystallization using synthetic templates': *Advanced Materials*, 'Crystallization in Patterns: A Bio-inspired Approach', 3 August 2004, pp. 1295–302.

p. 154 'The broad outline of how they are formed': Thompson (1961), pp. 154–69.

p. 154 'On the shore his pockets were filled with shells': Thompson (1958), p. 184.

p. 155 'To form a sphere from hexagons': Edmondson (1992), pp. 232–43.

p. 156 'A detergent template that could organize the formation of stable spheres': *Nature*, 'Lamellar aluminophosphates with surface patterns that mimic diatom and radiolarian micro-skeletons', 2 November 1995, pp. 47–50.

p. 156 'Producing radiolarian-like structures by letting mineralization of calcium carbonate occur at the boundaries of micro-emulsions': *Nature*, 'Fabrication of hollow porous shells of calcium carbonate from self-organizing media', 28 September 1995, pp. 320–3.

p. 158 'The collapse of Moore's Law': *Horizon*, BBC TV, 5 February 2004.

p. 158 'Nanorobots, created in atom-by-atom assemblers': Drexler (1986).

p. 158 'There wasn't much difference between creating a new bacteria to spew out, say, insulin molecules': Crichton (2002), p. 193.

Chapter 7

p. 161 'Would all the philosophers and all the armies in the world be able to construct this little fly': Levi, (2000), p. 19.

p. 161 'Less computational power than a toaster': *Nature*, 'Red admiral agility', 12 December 2002, p. 616.

p. 162 'In your class you seem to have talked about geese and swans': Tennekes (1996), p. ix.

p. 162 'The US Defense Advanced Research Projects Agency's (Darpa) $35 million development programme': a lengthy rationale for the Darpa programme, dating from 1997, is given at: http://www.darpa.mil/tto/ mav/mav_auvsi.html

p. 163 'To fly, wings that are moving horizontally': Vogel (1994), pp. 230–4.

p. 164 'So small creatures fall through the air slowly': Vogel (1999), pp. 44–5.

p. 164 'The patterns made in the air by an insect's wings': *Scientific American*, 'Solving the Mystery of Insect Flight', June 2001, pp. 49–57.

p. 166 'A breakthrough study of tethered hawkmoths': *Nature*, 'Leading-edge vortices in insect flight', 19/26 December 1996, pp. 626–30.

p. 166 'Standard aeroplane theory could not account for the lift generated by insects': *Philosophical Transactions of the Royal Society of London B*, 24 February 1984, pp. 1–181.

p. 167 'A model fruitfly 100 times life size': *Scientific American*, 'Solving the Mystery of Insect Flight', June 2001, pp. 49–50.

p. 168 'Studies on free-flying red admiral butterflies': *Nature*, 'Unconventional lift-generating mechanisms in free-flying butterflies': 12 December 2002, pp. 660–4.

p. 168 'When Darpa held a competition in 1999': a paper on the Darpa winner, the Black Widow, can be found at: http://www.aerovironment.com/area-aircraft/prod-serv/bwid pap.pdf

p. 170 'Ron Fearing began work on the MAV': Micromechanical Flying Insect (MFI) Project, http://robotics.eecs.berkeley.edu/#ronfMFI/mfi/html

p. 174 'The Eurofighter and the F22 have deliberately unstable flight dynamics': http://www.flug-revue.rotor.com/FRheft/FRH9909/FR9909e.htm.

p. 177 'The basic theory of this kind of motion detection': *Autonomous Robots*, 'A robust analog VLSI motion sensor based on the visual system of the fly', vol. 7, 1999, pp. 211–24.

p. 177 'Aeroplanes have few sensors while insects have hundreds of them': *IEEE Measurement & Instrumentation Magazine*, 'Sensor rich feedback control', September 2004, pp. 19–26.

p. 178 *fn* 'A challenge competition for wheeled robots': *Guardian*, 'Life' 18 March 2004, p. 2.

Chapter 8

p. 183 'The same way that some leaves, especially those of hornbeam and beech, unfold from the bud': *Proceedings of the Royal Society of London B*, 'The geometry of unfolding tree leaves', 18 September 1997, pp. 147–54.

p. 185 'The map idea was launched at the 10th Conference of International Cartographers': *British Origami*, 'The Miura-ori map', vol. 88, 1981, pp. 3–5.

p. 188 'In 1985, he took out a patent on the one-pull map': British Patent 2,173,448A.

p. 189 'Several "leaves" radiate to create a circular folding mechanism': *Philosophical Transactions of the Royal Society of London A*, 'Deployable membranes designed from folding tree leaves', 15 February 2002, pp. 227–38. This issue of the Royal Society's journal is devoted to bio-inspiration.

p. 189 'Solar panel arrays for space satellites': *Space Solar Power Review*, vol. 5, no. 4, 1985, pp. 345–56.

p. 189 'Solar-sail propulsion units for spacecraft': 'Inextensional wrapping of flat membranes'. In *Proceedings of the 1st International Seminar on Structural Morphology* (ed. R Motro & T Wester), 1992, pp. 203–15.

p. 190 'Bio-inspired camouflage': *EPSRC Newsline*, 'Nature's Secrets', Issue 20, Winter 2001, pp. 12–13.

p. 190 'A cylinder with these diamond patterns already built in': Proceedings of IASS Symposium on Folded Plates and Prismatic

Structures, International Association for Shell Structures, Vienna, 1970.

p. 191 'How honeycomb patterns came to be used in aircraft construction': Gordon (1991), pp. 297–9.

p. 193 'An out-of-plane triangular kink': *Philosophical Transactions of the Royal Society of London B*, 29 September 2003, p. 1579.

p. 194 'The single-celled organism *Vorticella*': *Science*, 'Motility Powered by Supramolecular Springs and Ratchets', 7 April 2000, pp. 95–9.

Chapter 9

p. 197 'The *Zoomorphic* exhibition at the Victoria and Albert Museum': Aldersey-Williams (2003).

p. 198 'Air-conditioning systems perfected by some species of African termites': *Natural History*, 'A superorganism's fuzzy boundaries', July–August 2002.

p. 198 'In the traditional Iraqi house, roof-top vents called *badgirs* draw cool air': http://www.brainworker.ch/Irak/architecture.htm

p. 198 'Pearce McComish's Eastgate Building in Harare': http://www.pearcemccomish.com/eastgate.htm

p. 199 'Norman Foster's Swiss Re Tower': Aldersey-Williams (2003), pp. 98–100.

p. 199 'D'Arcy Thompson wrote about early links between building and nature': Thompson (1961), pp. 19–22.

p. 199 'Paxton was well-placed to be bio-inspired': Colquhoun (2003). Paxton's latest biographer describes well the dizzying achievements of this workaholic son of a farm labourer who began as a gardener's apprentice and became one of Victorian England's great movers and shakers.

p. 200 'Paxton gave a talk to the Royal Society of Arts': *The Times*, 14 November 1850.

p. 201 'The arch is the best solution': Gordon (1991), p.171–97. Gordon is brilliant on the principle of building with compression and tension – from the Greek temples through the buttresses of medieval cathedrals to the tension fields in the skin of aircraft.

p. 202 'A cantilever bridge, a structure engineers share with nature': Thompson (1961), pp. 241–58.

p. 203 'The tubes of the bridge are stiffened': Thompson (1961), pp. 229–30.
p. 203 'Enough hands, enough grip': Mayakovsky (1966).
p. 203 'Hart Crane's 1933 epic of the United States': Crane (1992), p. 2.
p. 204 'Reinforced concrete was invented by a Frenchman': *see* Billington (1985), pp. 148–51. Billington's book on structural engineering as an artform makes a vital link between architectural form finding and principles of natural beauty.
p. 204 'A later development, pre-stressed concrete': Billington (1985), pp. 194–212.
p. 204 'The great Swiss bridge-builder Robert Maillart': Billington (1997).
p. 204 'Memorable buildings in concrete': Billington (1985), pp. 171–93.
p. 204 'The Swiss engineer Heinz Isler': Chilton (2000).
p. 208 The essence of beauty of line in painting': Thompson (1961), pp. 287–8.
p. 208 *fn* 'A graded lessening': Hogarth (1971), p. 49 and plate 1, fig. 49. Hogarth says: 'There is but one precise line, properly to be called the line of beauty.' This line, like an elegantly flattened S, is more or less the curve of a recumbent human back from the shoulders to the buttocks; it provided the title for Alan Hollinghurst's Booker prize-winning novel *The Line of Beauty* (Picador, 2004).
p. 208 '"That's my crane!", he exclaimed': Thompson (1961), pp. 230–3.
p. 209 'In Switzerland they had brotherly love': Greene (1998), p. 100. The famous lines were not in Greene's original script and were added in production, probably by Orson Welles.
p. 210 'He said that it was as if he had been climbing a wall': Chilton (2000), p. 23.
p. 211 'When on a cold winter night': Chilton (2000), p. 149.
p. 213 'Santiago Calatrava is currently the leading exponent of organic architecture': Jodido (1998).
p. 214 'The Dinosaur Bridge': Thompson (1961), p. 257.
p. 215 'The Dinosaur Bridge won a 1988 competition': *New Civil Engineer*, 21 January 1988, pp. 27–9.

p. 216 'Geodesic domes, created from complex webs of triangular or hexagonal and pentagonal units': Edmondson (1992), pp. 232–43.

p. 217 'The biologist Aaron Klug discovered the icosahedral structure of phages in 1962': *Cold Spring Harbor Symp. Quant. Biol.*, 'Physical Principles in the Construction of Regular Viruses', vol. 27, 1962, pp. 1–24.

p. 217 'A geodesic dome on an even smaller scale': Aldersey-Williams (1995).

p. 218 'The Eden Project': Smit (2002).

p. 218 'The best structures': http://www.edenproject.com/3440_3459.htm

p. 218 'Where bubbles intersect they do so vertically in a plane': http://www.edenproject.com/3440_3459.htm

p. 220 'Our bodies provide a familiar example of a pre-stressed tensegrity structure': *The Journal of Cell Science,* 'Tensegrity I. Cell structure and hierarchical systems biology', 1 April 2003, pp. 1157–73.

p. 220 'Its genesis was muddled': *International Journal of Space Structures*, 'Origins of Tensegrity: Views of Emmerich, Fuller and Snelson', vol. 11, nos 1 and 2, 1996, pp. 43–52.

p. 222 'Geiger developed his tensegrity cabledome system in 1984': Robbin (1996), pp. 33–7. Robbin's book, whilst not explicitly concerned with bio-inspiration, is the best guide available to advanced engineering building structures based on geometric principles, economy of materials and new materials.

p. 225 'Frei Otto': Robbin (1996), pp. 12–13.

p. 228 'Donald Ingber takes tensegrity into the hidden world of the cell': *The Journal of Cell Science,* 'Tensegrity I. Cell structure and hierarchical systems biology', 1 April 2003, pp. 1157–73.

p. 228 '*Vorticella,* the tiny self-deploying organism': *Science*, 'Motility Powered by Supramolecular Springs and Ratchets', 7 April 2000, pp. 95–9.

p. 229 'Masts for space arrays using erectable tensegrity structures': Gunnar Tibert, Ph.D thesis, 2002, Royal Institute of Technology, Stockholm, Sweden. A video of the mast deployment is available at: http://www-civ.eng.cam.ac.uk/dsl/research/agt27/TriplexMast2.mpg

Chapter 10

p. 231 'The *De Rerum Natura* of Lucretius': Calvino (1992), p. 8. Calvino was one of a long line of Italians – including Lucretius, Leonardo, Galileo and Primo Levi – for whom there was only one culture. Each was primarily an artist or a scientist but in every case the other discipline was central to their thought.

p. 232 'To understand the mechanism of nature does not detract from the poetry': Dawkins (1998).

p. 234 'That nature keeps some of her secrets': Thompson (1961), p. 171.

FURTHER READING

Aldersey-Williams, Hugh, *The Most Beautiful Molecule: The Discovery of the Buckyball*, Aurum, London, 1995

Aldersey-Williams, Hugh, *Zoomorphic: New Animal Architecture*, Lawrence King Publishing, London, 2003

Asquith, RS, *Chemistry of Natural Protein Fibres*, Plenum Press, New York and London, 1977

Bachelard, Gaston, *The Poetics of Space* (trans. Maria Jolas), Beacon Press, Boston, 1969

Baeuerlein, Edmund, (ed.), *Biomineralization*, Wiley-VCH, Weinheim, 2000

Baricco, Alessandro, *Silk* (trans. Guido Waldman), Harvill, London, 1997

Barthlott, Wilhelm, 'Scanning electron microscopy of the epidermal surface in plants', in D Claugher (ed.), *Scanning Electron Microscopy in Taxonomy and Functional Morphology*, Oxford University Press, 1990

Barthlott, Wilhelm et al., 'Nelumbo: Biology and Systematics of an Exceptional Plant', in Z Aoluo and W Sugong (ed.), *Floristic Characteristics and Diversity of East Asian Plants*, Springer Verlag, Berlin, 1998

Billington, David P, *The Tower and the Bridge*, Basic Books, New York, 1985

Billington, David P, *Robert Maillart*, Cambridge University Press, 1997

Birks, JB, *Rutherford at Manchester*, Heywood & Company Ltd, London, 1962

Bon, Comte Xavier Saint-Hilaire, *Dissertation sur l'utilité de la soie des araignées*, Avignon, 1748

Cairns-Smith, AG, *Seven Clues to the Origin of Life*, Cambridge University Press, 1990

Calvino, Italo, *Six Memos for the Next Millennium*, Jonathan Cape, London, 1992

Chilton, John, *Heinz Isler*, Thomas Telford Ltd, London, 2000

Claugher, D (ed.), *Scanning Electron Microscopy in Taxonomy and Functional Morphology*, Oxford University Press, 1990

Colquhoun, Kate, *A Thing in Disguise: The Visionary Life of Joseph Paxton*, Fourth Estate, London, 2003

Conard, Henry Shoemaker and Hus, Henri, *Water-Lilies and How to Grow Them*, Doubleday & Co., London, 1907

Cott, HB, *Adaptive Coloration in Animals*, Methuen, London, 1957

Crane, Hart, *The Bridge*, Liveright, New York and London, 1992

Crichton, Michael, *Prey*, HarperCollins, London, 2002

D'Abrera, Bernard, *Butterflies of the neotropic region, Part II*, Hill House, Ferny Creek, Victoria, Australia, 1984

Dawkins, Richard, *Climbing Mount Improbable*, Viking, London, 1996

Dawkins, Richard, *Unweaving the Rainbow: Science, Delusion and the Appetite for Wonder*, Allen Lane, London, 1998

Drexler, Eric K, *Engines of Creation*, Fourth Estate, London, 1986

Edmondson, Amy C, *A Fuller Explanation: The Synergetic Geometry of R. Buckminster Fuller*, Van Nostrand Reinhold, New York, 1992

Feynman, Richard, *The Pleasure of Finding Things Out: The Best Short Works of Richard Feynman*, Penguin Books, London, 2001

Foelix, RF, *Biology of Spiders*, Oxford University Press, 1996

Fox, HM and Vevers, Gwynne, *The Nature of Animal Colours*, Sidgwick and Jackson, London, 1960

Fujishima, Akira et al., *TiO_2 Photocatalysis – Fundamentals and Applications*, BKC Inc., Tokyo, 1999

Gordon, JE, *Structures: or Why Things Don't Fall Down*, Penguin Books, London, 1991

Greene, Graham, *The Third Man*, Heinemann, London, 1976

Hardy, A, *The Open Sea: The World of Plankton*, Collins, London, 1956

Harvey, EW, *Bioluminescence*, Academic Press, London, 1952

Hillyard, Paul, *The Book of the Spider: From Arachnophobia to the Love of Spiders*, Hutchinson, London, 1994

Hogarth, William, *The Analysis of Beauty, 1753*, Scolar Press, Menston, 1971

Holbrook, David, *Selected Poems 1961–1978*, Anvil Press, London, 1980

Holub, Miroslav, *Poems: Before & After*, Bloodaxe Books, Newcastle, 1990

Hongu, T and Phillips, GO, *New Fibres*, 2nd edn, Woodhead Publishing, Cambridge, 1997

Hooke, Robert, *Micrographia*, 1665; reprinted in RT Gunther, *Early Science in Oxford, Vol. XIII, The Life and Work of Robert Hooke (Part V)*, Oxford University Press, 1938

Joannopoulos, John D et al., *Photonic Crystals: Molding the Flow of Light*, Princeton University Press, Princetown, NJ, 1995

Jodido, Philip, *Santiago Calatrava*, Taschen, Cologne, 1998

Jones, Richard AL, *Soft Machines*, Oxford University Press, 2004

Kaplan, David et al., *Silk Polymers, Materials Science and Biotechnology*, ACS Symposium Series 544, American Chemical Society, Washington DC, 1994

Kato, Bunno et al. (trans.), *The Threefold Lotus Sutra*, John Weatherill Inc., New York, 1975

Komai, Gonnoské, in *Lotus and Chrysanthemum*, Joseph Lewis French (ed.), Boni & Liveright, New York, 1927

Leigh, Richard, *Poems*, 1675, reprinted with an introduction by Hugh Macdonald, Blackwell, Oxford, 1947

Levi, Primo, *Other People's Trades*, Michael Joseph, London, 1989

Levi, Primo, *The Periodic Table*, Penguin Books, London, 2000

Lucretius, *The Way Things Are: The De Rerum Natura of Titus Lucretius Caro* (trans. Rolfe Humphries), Indiana University Press, 1969

Mann, Stephen, *Biomineralization: Principles and Concepts in Bioinorganic Materials Chemistry*, Oxford University Press, 2001

Mann, Thomas, *Dr Faustus* (trans. John E Woods), Vintage International, New York, 1999

Mayakovsky, Vladimir, 'Brooklyn Bridge' (trans. Vladimir Markov and Merrill Sparks), from *Modern Russian Poetry*, MacGibbon & Kee, London, 1966

Nabokov, Vladimir, *Speak, Memory*, Penguin Books, London, 2000

Nachtigall, Werner, *Biological Mechanisms of Attachment*, Springer, Berlin, 1974

Nealson, Kenneth H, *Bioluminescence: Current Perspectives*, Burgess Publishing Co., Minneapolis, 1981

Newton, Sir Isaac, *Opticks*, reprinted from the 4th edition, G Bell & Sons, London, 1931

Niklos, David A and Freyer, Greg A, *DNA Science: A First Course*, 2nd edition, Cold Spring Harbor Laboratory Press, New York, 2003

Parker, Andrew, *In the Blink of an Eye*, The Free Press, London, 2003

Pianka, Eric R and Vitt, Laurie J, *Lizards: Windows to the Evolution of Diversity*, University of California Press, Berkeley, 2003

Pratt, EA, *Two Years Among New Guinea Cannibals*, Lippincott, Philadelphia, 1906

Ramsay, RA, *The Experimental Basis of Modern Biology*, Cambridge University Press, 1965

Robbin, Tony, *Engineering a New Architecture*, Yale University Press, New Haven and London, 1996

Silverman, Mark P, *Waves and Grains: Reflections on Light and Learning*, Princeton University Press, Princeton, NJ, 1996

Smit, Tim, *Eden*, Corgi, London, 2002

Stevenson, Anne, *Collected Poems*, Bloodaxe Books, Newcastle, 2000

Tennekes, Henk, *The Simple Science of Flight*, The MIT Press, Cambridge, MA, 1996

Thompson, D'Arcy Wentworth, *On Growth and Form*, abridged edn, JT Bonner (ed.), Cambridge University Press, 1961

Thompson, Ruth D'Arcy, *D'Arcy Wentworth Thompson, the scholar-naturalist, 1860–1948*, Oxford University Press, 1958

Van Dulken, Stephen, *Inventing the 20th Century*, British Library, London, 2000

Van Renesse (ed.), *Optical Document Security*, 2nd edn, Artech House, Boston and London, 1998

Vogel, Steven, *Life in Moving Fluids: The Physical Biology of Flow*, 2nd edn., Princeton University Press, Princeton, NJ, 1994

Vogel, Steven, *Cats' Paws and Catapults*, Penguin Books, London, 1999

Von Frisch, Otto, *Animal Camouflage*, Collins, London, 1973

Vukusic, Pete, 'Natural Coatings', in Norbert Kaiser and Hans K Pulker (eds), *Optical Interference Coatings*, Springer, Berlin, 2003

INDEX

LOWER MERION LIBRARY SYSTEM
30043004888156